The Engaged Sociologist

Sixth Edition

This book is dedicated to two generations that inspire us.
First, to the generation that came before us, most particularly
our parents, Ilene, Paul, Marnie, and Joe, who have always encouraged us
to ask questions, and who embrace our full selves. Second, to the generation
that comes after us, most particularly the young activists who courageously
challenge inequitable systems, continually invigorate movements for social
change, and work tirelessly for a more socially just world.

The Engaged Sociologist

Connecting the Classroom to the Community

Sixth Edition

Jonathan M. White

Bentley University

Shelley K. White

Simmons University

Los Angeles | London | New Delhi
Singapore | Washington DC | Melbourne

FOR INFORMATION:

SAGE Publications, Inc.
2455 Teller Road
Thousand Oaks, California 91320
E-mail: order@sagepub.com

SAGE Publications Ltd.
1 Oliver's Yard
55 City Road
London EC1Y 1SP
United Kingdom

SAGE Publications India Pvt. Ltd.
B 1/I 1 Mohan Cooperative Industrial Area
Mathura Road, New Delhi 110 044
India

SAGE Publications Asia-Pacific Pte. Ltd.
18 Cross Street #10-10/11/12
China Square Central
Singapore 048423

Library of Congress Cataloging-in-Publication Data

Names: White, Jonathan M. (Jonathan Michael), author

Title: The engaged sociologist : connecting the classroom to the community / Jonathan M. White, Bentley University, Shelley K. White, Simmons University.

Description: Sixth Edition. | Thousand Oaks, Calif. : SAGE, [2019] | Previous edition: 2015. | Includes bibliographical references and index.

Identifiers: LCCN 2019007141 | ISBN 9781506347462 (Paperback : acid-free paper)

Subjects: LCSH: Service learning—United States. | Sociology—Study and teaching (Higher)—United States.

Classification: LCC LC221 .K68 2019 | DDC 361.3/70973—dc23 LC record available at https://lccn.loc.gov/2019007141

This book is printed on acid-free paper.

Acquisitions Editor: Jeff Lasser
Editorial Assistant: Tiara Beatty
Production Editor: Gagan Mahindra
Copy Editor: Karin Rathert
Typesetter: C&M Digitals (P) Ltd.
Proofreader: Dennis Webb
Indexer: Mary Mortensen
Cover Designer: Alexa Turner
Marketing Manager: Will Walter

SUSTAINABLE FORESTRY INITIATIVE

Certified Chain of Custody
At Least 10% Certified Forest Content
www.sfiprogram.org
SFI-01028

19 20 21 22 23 10 9 8 7 6 5 4 3 2 1

Contents

Preface

A Note From the Authors to the Students Reading This Book

S ociology is the coolest academic discipline. Seriously, what other area of study is better at helping you figure out how society operates and how you can use that knowledge to create social change? Both of us were drawn to sociology because we wanted to figure out how to fight injustice and promote democracy more effectively. It has also guided us in everyday life tasks, such as figuring out how to get policies passed on campus, deciding whom to vote for, and learning why it's vital to earn a college degree in a service-based economy. This book is part of our efforts to get students hooked on sociology and, in the process, help them to become engaged and effective citizens who can strengthen our democratic society.

This book is also part of a larger, national effort to "educate citizens" by encouraging students to participate in civic engagement exercises that connect the classroom to the community. Organizations such as The Democracy Imperative, Campus Compact, and the American Democracy Project are establishing movements to make civic engagement a part of the college experience for all undergraduates. College leaders all across the country realize that as educators they are obligated to give students the tools they will require to be effective citizens as well as the skills they will need in the workforce. Leaders in *all* sectors of society understand that higher education, when connected to the larger society, benefits everyone, on *and* off campus.

We believe, as leaders of the American Sociological Association have noted when promoting public sociology, that sociology is particularly suited to teaching students what they need to know to become effective and full members of our society. As the prominent sociologist Randall Collins has pointed out, the two core commitments of sociology are (a) to understand how society works and (b) to use that knowledge to make society better. We believe that helping students learn how to think sociologically and use sociological tools is, in effect, enabling them to become better citizens. No doubt, the professors who assigned this book to you also share this belief. They will gladly tell you why *they* think sociology is an incredibly useful and practical academic discipline.

We also know that sociology is fun to learn and to teach. That's why we created a book that we hope will be enjoyable to use for both students and teachers. The exercises throughout the chapters allow students to connect the sociological knowledge that they are learning to their campus and the larger community. So as soon as you develop your sociological eye, you will make use of it! *Please note that you will need to make sure you follow the rules for research on human subjects and get approval from the Institutional Review Board on your campus before carrying out some of these exercises.* (Your professor will tell you how to do so.) This book will also help you to connect your own life to the larger society as you learn about the "sociological imagination" and the power it has to positively affect your community. The Sociologist in Action sections in each chapter will give you powerful examples of how sociology students and professional sociologists (both professors and applied sociologists) use sociology in myriad ways in efforts to improve society. By the end of the book, you can create your own Sociologist in Action section, in which you'll show how you used sociological tools in efforts to influence society. If you want to see more examples of sociologists in action, please check out our other books, *Sociologists in Action: Sociology, Social Change, and Social Justice* and *Sociologists in Action on Inequalities: Race, Class, Gender, and Sexuality.*

We look forward to seeing your Sociologist in Action pieces and featuring many of them in future editions of this book and on the website for *The Engaged Sociologist* (http://study.sagepub.com/white6e). In the meantime, we hope you enjoy the book and use the knowledge and skills you gain from it to make yourself a more effective citizen, strengthen our democracy, and work for a more just and civil society. We think that you will discover what we discovered when we began our journey as sociologists—that sociology is a cool and powerful tool. And, of course, we hope that you have a lot of fun in the process!

Acknowledgments

This sixth edition would not have been possible without the very good assistance of many people. We would first like to thank our very supportive editor, Jeff Lasser. Jeff has brought energy, flexibility, and great support to our writing of the book, and we are indebted to him and truly grateful. We feel lucky to work with him and to benefit from his great insight and encouragement. We would also like to thank our former editors: Dave Repetto, a dear friend, who brought a commitment to connecting the classroom to the community, wonderful insight, unbending support, and a wealth of good judgment to this project, and Ben Penner, our original editor, for his excellent work and for believing in the vision of this book. Ben's enthusiasm for *The Engaged Sociologist* and for public sociology in general inspired us as we conceptualized and wrote this book. We also are indebted to Gagan Mahindra, our production editor, for his strong support and ability to guide us throughout the course of this project. We would particularly like to thank Karin Rathert for her exceptionally astute and thoughtful work copyediting the book, it has been a pleasure to work with her. Tiara Beatty, our steadfast editorial assistant, has done a wonderful job keeping things organized and on time and responding to all of our requests and needs. We also owe a special debt of gratitude to our colleague, Howard Lune, for his early contributions to this book. And a very deep debt of gratitude to Brian Shea, who offered tremendous and unwavering help and insight on this edition. Finally, we have deep gratitude for Kathleen Korgen, co-author of editions 1 through 5 of this book. Kathleen's vision, engaging style of writing for undergraduate students, commitment to social change, and steadfast work ethic have been a guiding force and remain core to the DNA of this book.

Jonathan and Shelley would like to thank the many students who have inspired us with their energy, intellect, and unwavering belief that social change is indeed possible. Jonathan would like to thank the amazing student leaders and the dedicated and skilled staff of the Bentley Service-Learning and Civic Engagement center and his former student leaders of the Social Justice League and the Human Rights Action Committee. Shelley would like to thank the student activist leaders of HERE4Justice, the Social Justice Coalition, and Students United for Justice, as well as her colleagues in health equity who inspire her every day. Together, we thank the extraordinary staff of WE and ME to WE and the millions of young people involved in their programs, working tirelessly for a better world. A huge thanks also goes to our family, who provide us with unwavering support, love, and happiness and who inspire us to continuously work toward creating a stronger civil society. And a very special group of people, our 13 nephews and nieces, deserve a special debt of gratitude because they are particularly inspirational to us in so many ways!

List of Reviewers

Beverly Bennett, Wright College

James N. Maples, Eastern Kentucky University

Aloma Mendoza, Saint Leo University

CHAPTER 1

The Engaged Sociologist

The Sociological Perspective and the Connections Among Sociology, Democracy, and Civic Engagement

Have you ever wanted to help change society? Do you want to have a voice in how things work throughout your life? If so, you've come to the right discipline. Sociology helps you to understand how society operates and, in turn, how to make society better.

Sociology is the scientific study of society. As sociologists, we see how individuals both shape and are shaped by larger social forces. By developing what is called a *sociological eye* (Collins, 1998; Hughes, 1971), we are able to look beneath the surface of society and see how it really works. For example, with a sociological eye, we can recognize the tremendous influence of culture on individuals. Imagine how different you might be if you had grown up in Sweden, Ethiopia, Bangladesh, or another country with a culture very different from your own.[1] You would still look about the same (though you'd have different mannerisms, speak a different language, and have a different haircut and clothes), but your values, norms, and beliefs would be different. Your view of the proper roles of men and women, your religious or secular values, your career goals, your views about race, class, and sexuality, your education, and so forth, are shaped by the society in which you grew up.

Look at the differences between your immediate family and some of your relatives who may have much more or much less money. Does social class cause the differences, or to what extent do the differences help determine the social class to which we will belong?[2] Consider the varying perspectives that your male relatives and your female relatives bring to the same questions. They all live in the same world, in close proximity perhaps, but they have had such different experiences of it that some people even joke that men and women come from different planets (Gray, 2004). Now, consider the people in your life who may not identify as male or female and the perspectives and life experiences they bring. By using the sociological eye, we can look at the world from a unique angle, notice what is often unobserved, and make connections among the patterns in everyday events that the average person might not notice. In doing so, we can understand how different organizations, institutions, and societies function; how social forces shape individual lives and ideas; and, in turn, how individuals shape organizations and institutions.

By viewing society through the perspective of a *social world model*, we can study different levels of social units—from small- to large-scale parts of the world—that interact with one another. For example, we can study participation in the democratic process through examining *interpersonal and local organizations* (e.g., political activism among students on your campus and your school's College Democrats and College Republicans clubs), *larger organizations and institutions* (e.g., national Democratic and Republican Parties and state boards of elections), and *nations or global communities* (e.g., the U.S. presidential electoral process and the UN Sustainable Development Agenda implementation). No matter which of these we might choose to study, however, we would always be particularly concerned with the connections between the varying social groups. For example, using the social world model would help us see that individuals are influenced by and can influence their classmates, their political party, their nation, and their global community.

The social world model and a sociological eye also enable us to recognize persistent patterns that work to create disadvantages for certain groups in society, resulting in institutional discrimination (intentional or unintentional structural biases). For example, U.S. society functioned in such a way for over 200 years that there were no female Supreme Court justices before President Ronald Reagan appointed Sandra Day O'Connor in 1981. Sociologists, using the sociological eye, recognize that the long-standing all-male makeup of the Supreme Court was part of a larger pattern of sex discrimination. Some of the discrimination was deliberate and based on people's ideas about gender. Some of it was political, based on a calculation of how the public would respond to the nomination of a woman to such a post. Some of it even had to do with the fact that our culture tends to use similar language and ideas to describe both *leadership qualities* and *masculinity*. Thus, when people think of leadership, they tend to associate it with the qualities that men often bring to the table (Schein, 2001).[3] Social science research on the connections among gender roles, socialization, and sex discrimination, such as Betty Friedan's book *The Feminine Mystique* (1963), which shattered the myth that women could find fulfillment only as wives and homemakers, became part of public knowledge and was used to make the case for the women's movement in the 1960s and 1970s. Ultimately, with many other feminist scholars contributing research, this movement paved the way for a political environment conducive to the appointment of Sandra Day O'Connor and, eventually, Ruth Bader Ginsburg, Sonia Sotomayor, and Elena Kagan.

The social world model and a sociological eye are useful tools in exploring a gamut of social patterns and outcomes. As another example, same sex marriage has been against both the law and general moral viewpoint of society for most of U.S. history. Even as recently as 2001, 57 percent of Americans opposed gay marriage and just 35 percent supported it. However, the most recent Pew survey in 2017 indicates that there has been a strong change in how Americans view this issue, flipping their beliefs in just a decade and a half. Today, 62 percent of Americans, including 74 percent of millennials, support gay marriage while only 32 percent oppose it (Pew Research Center, 2017). In 2004, Massachusetts became the first state in the country to legalize gay marriage (Rudick, 2004). Over the following decade, many states followed by either passing laws to legalize same sex marriage or passing laws to ban it, eventually creating court battles that led to the issue being heard by the Supreme Court. On June 26, 2015, the Supreme Court handed down a landmark decision in the *Obergefell v. Hodges* case, stating that it is illegal for states to ban same sex marriage and thus legalizing it (CNN, 2018).

Use of the sociological eye also helps efforts to persuade governmental officeholders to initiate social policies that address social inequities. By looking beneath the surface of governmental operations, we can answer questions such as the following: To whom do office holders tend to respond the most? Why? How can we use this information to make sure that they respond to us? What social forces compelled Ronald Reagan, one of our most conservative presidents and not known as a women's rights advocate, to become the first president to choose a female Supreme Court justice? What explains the rapid change in public opinion regarding same sex marriage? How might public opinion have influenced policy and legal decisions? What does this all tell us about the ability to create social change, even on issues that are seemingly set in stone?

The Two Core Commitments of Sociology

According to Randall Collins (1998), using the sociological eye is one of two "core commitments" of sociology. The second is *social activism*. Once we understand how society operates, we are obligated to participate actively in efforts to improve it.

The sociological eye and social activism go hand in hand. The sociological eye helps us to uncover social patterns in society, and our use of those findings leads us to become effective, engaged citizens. Once you start seeing the world through a sociological eye, you will see social patterns all around. Through the use of sociological research methods, you can also learn how to discover and analyze more social patterns and even to propose solutions to social problems based on the results your research has produced. The more you train yourself as a sociologist, the stronger your sociological eye and ability to practice effective and constructive social activism will become.

And if you think you're too young or don't have the power to work toward social change, the recent protests by the Dakota Sioux over the 1,168 mile Keystone XL oil pipeline, moving through four states, is one example that teaches otherwise. Some politicians and business leaders touted the pipeline as important for economic growth and for the United States to gain less dependency on other nations for our oil. Many others, led by the Dakota Sioux, protested the damage the pipeline will create: "The construction and operation of the pipeline . . . threatens the Tribe's environmental and economic well-being, and would damage and destroy sites of great historic, religious, and cultural significance to the Tribe" (Park, 2016). Additionally, the tribe's lawsuit put forward that digging under the Missouri River for the pipeline would pollute their water supply. Many youth activists were at the forefront of activism and advocacy to stop the pipeline, from taking direct action such as putting their bodies in the way of machinery, to writing letters to key government and business leaders, to even suing the government for impinging on their future by polluting their land and destroying its viability.

You will see throughout this book how sociology can be used to make a positive impact on society. Each chapter contains examples of *sociologists in action*, those who have honed their sociological eye and use their sociological skills to create social change. In doing so, they fulfill the two commitments of sociology.

The Sociological Imagination

To understand how we might influence society, we must first understand how we are affected by it. C. Wright Mills (1959/2000) described this ability as the "sociological imagination." When we begin to relate personal troubles to public issues, connecting our individual lives to what's happening in our society, we are using the sociological imagination.

For instance, one of us experienced our parents divorcing. As an individual, this was a personal trouble for him. Using the sociological imagination, we see that he was a part of a cohort of U.S. children who lived through the great rise in the divorce rate in the 1970s, a decade when the rate nearly doubled from the prior decade. If he had been a child during the 1870s, his parents would most likely have remained married. However, the changes that our society went through in the 1960s and 1970s (legal rights and protections for women, which enabled more women to exit relationships; the decline of religiosity; cost of living increases, which required more women to join the workforce; etc.) resulted in an increase in the divorce rate and, in turn, his own parents' divorce. His personal trouble (the divorce of his parents) was directly related to a public issue (the society-wide rise in the divorce rate).

Today, both of us struggle to buy clothes for our young nieces that do not resemble those of the women on *The Real Housewives* and *Pretty Little Liars* shows. As an aunt and an uncle, we are horrified that anyone would expect young girls to wear such skimpy outfits (particularly *our* nieces!). As sociologists, we can look at a sample of clothing stores and advertisements in the United States and quickly realize that our experience is part of a society-wide pattern of sexualizing girls—even very young girls (Healy, 2012). Sociologists and psychologists have produced significant research, illustrating that "the toxic mix of sexualizing media and commodities (e.g., Bratz dolls, thongs, tee-shirts) transforms girls between the ages of 8 and 12 (or "tweens") into self-sexualizing subjects at risk for a host of mental, physical, cognitive, and relational problems" (Egan, 2013, p. 75).

Given that these are the outcomes for girls, we might then want to start to research why a society like ours, with such a long history of public activism around "standards of moral decency," is so consistent—almost aggressive—in the sexualizing of girls. One hypothesis we might test, for instance, is whether this social behavior is related to the relative absence of mothers in the highest positions of fashion design and marketing. If we were to discover this to be true, we could use our findings to work for social change, trying to make these workplaces more open and inviting for fashion designers and marketers who are mothers.

One of the functions of sociology, as Mills (1959/2000) defined it, should be to "translate personal troubles into public issues" (p. 187). Once you start using your sociological imagination and looking at the world through the sociological eye and the social world model, it's impossible not to notice the connections between yourself as an individual and larger societal patterns. Consider the kind of job you hope to have after leaving college. Will you make an annual salary or an hourly wage? Will you be able to support a family? Will your job help you make a meaningful life for yourself? Will you have job security? Will your position lead to movement up or down the social class ladder? How will your job compare with those in the corner office of a major corporation or those found on the floor of a sweatshop? Will your job allow you to make the positive impact you want to on society and to create the social change you believe to be necessary?

Sweatshops are production sites where workers face near-slavery conditions, with arbitrary punishments and few or no protections from unsafe work environments, and where they work for below living-wage pay. Sweatshop jobs do not come with insurance, sick days, retirement plans, or protection against arbitrary termination. These sites may seem very distant from your life and from the lives of most college students. On the surface, colleges and sweatshops seemingly have nothing to do with each other. However, if you look underneath the surface (or, perhaps, at what you or your classmates are wearing), you may see a connection. The students at Duke University did: When they learned of the horrible sweatshop conditions in which most of their Duke-labeled clothes were being manufactured, they mobilized and established a United Students Against Sweatshops (USAS) group on campus. Their efforts and those of several administrators at Duke (particularly the director of Duke Store Operations) sparked a campus-wide discussion about sweatshops and the university's responsibility to ensure that clothing with a Duke label is "sweat-free." In 1997, Duke was the first institution of higher education in the United States to adopt a code of conduct for mandating that the apparel companies with which they do business must submit to independent monitoring of the conditions in their factories. In 1999, Duke helped establish the Fair Labor Association (FLA), a collaboration of companies, civic organizations, and universities that began monitoring factories to root out abuses of workers. As of 2018, nearly 200 colleges and universities were affiliated with the FLA (FLA, 2018).

In 2000, another group of students, experts on labor rights, and university administrators established the Worker Rights Consortium (WRC) to assist in the enforcement of the codes of conduct established between colleges and universities and the companies that manufacture clothes for them. By 2017, 193 institutions of higher education and six high schools were WRC affiliates (WRC, 2018).[4] Today, students in universities around the country are creating Fair Trade Universities, working with their schools' administrators to "embed Fair Trade practices and principles into policy, as well as the social and intellectual foundations of their communities." There are 109 active Fair Trade University campaigns and 61 active Fair Trade Schools (K–12) campaigns across the country (Fair Trade Campaigns, 2018) (http://fairtradecampaigns.org/about/; go to http://fairtradecampaigns.org/campaign-type/universities to find out more about the Fair Trade Universities movement).

Sociology and the Critical Consumption of Information

In addition to having a trained sociological eye and making use of the sociological imagination, sociologists are informed and critical consumers of the barrage of information coming at us from all directions. Sociological research methods guide how we conduct research and how we interpret the information relayed by others. By understanding how "good" research is done, we can evaluate the information disseminated throughout society and know what news sources are trustworthy. These skills help us in our efforts to both understand and change society. In Chapter 3, we will outline in greater detail how sociological research methods can be used in this way.

Sociology and Democracy

Through reading this book and carrying out the exercises within it, you will learn how to look beneath the surface of social events, connect personal troubles to public issues, and know what information sources are trustworthy. You can then use these sociological tools to strengthen our society, make our nation more democratic, and work toward ensuring the rights and well-being of people all around the world. Although democracy is defined in different ways by a multitude of scholars, all point out that it is a system of governance that instills state power in citizenship rather than in government. This book shows how sociology can enable citizens (like you!) to become knowledgeable, active, and effective participants in our democratic society.

As you read the rest of this book, please think about what these sociology students have accomplished and how you too can use what you learn from this course to become a *sociologist in action!*

Sociologist in Action

The Social Justice League at Bridgewater State University

Joshua Warren, Bria Wilbur, Curtis Holland, and Jillian Micelli were four sociology majors at Bridgewater State University (BSU) who used their sociological eyes to make the connection between their campus and the community. Wanting to use the knowledge they learned in their courses to help effect positive social change, they created a student group called the Social Justice League. This group organized many events to educate the campus community about social justice issues, raise funds, and move the college toward more just practices.

One of the Social Justice League's big events was the creation of a "Tent City"—students, faculty, and staff slept outside in tents during a cold week in November. Each day, Tent City speakers from area shelters and organizations spoke to the campus community, and faculty from all across the campus brought their classes to these lectures. Students sleeping in Tent City were not allowed to use computers or cell phones, ate their meals in a "mock soup kitchen" set up in the cafeteria, and could bathe only by using public showers during set hours. In addition to creating the educational and symbolic components of Tent City, the students also raised several thousand dollars in cash and supplies to support a local homeless shelter.

(Continued)

(Continued)

The Social Justice League launched another strategic and powerful education and awareness campaign on campus when its members organized a series of educational events and demonstrations about sweatshop labor and the clothes sold at their college bookstore. They showed videos on the topic of sweatshops to the campus community, gave talks to classes, and staged a "mock sweatshop." Through these efforts, the students created a high level of campus awareness about sweatshops.

The Social Justice League also brought the issue of sweatshops to the direct attention of Dr. Dana Mohler-Faria, the president of BSU, who worked with the group to form a campus council to research the suppliers of the clothing sold at the student bookstore. After carefully examining the issue, the president agreed that BSU should join the WRC. Bria summed up the campaign by saying, "We worked extremely hard and I never felt more proud of myself than when I got the phone call that our campaign was successful and that BSU was going to join the Worker Rights Consortium! It was through sociology that I learned about these issues." Thanks to the efforts of the skilled and passionate members of the Social Justice League, BSU joined the growing ranks of colleges committed to ensuring that their campuses are sweat free.

Jillian described the connection between her sociological training and her actions to work for the betterment of society like this:

Within one year, my life changed dramatically. [By changing] my major to sociology the night before classes started, not only did I gain a degree in sociology, I also gained knowledge I would have never gained otherwise. The training and skills I learned from courses on genocide to courses on social inequality truly motivated me to strive for social change not only in the Social Justice League, but beyond. It taught me the connection I have with the global market and how if one pushes for something they believe in, social change will follow.

Speaking of his work as president of the Social Justice League, Joshua said,

By combining what I've learned in my studies with my passion for social justice and civic engagement, I am able to not only serve those immediately in need but also think about the reasons for why there is a need. It enables me to do more than just put a 'band-aid' on a problem. I can act proactively to find the source of the issue and uproot it!

Curtis made his awareness of the two core commitments of sociology clear when he said, "As I began to realize that I was becoming a burgeoning sociologist, I realized simultaneously that the knowledge I was gaining from sociological inquiry came with a great responsibility."

Recently, the students of the Social Justice League joined forces with students from the Free The Children Bridgewater State University chapter to work with administrators across the campus and help BSU become one of the first universities in the country to gain Fair Trade University status.

Exercise 1.1
How Is Higher Education Related to Democracy?

If you live in a democracy, then you have inherited certain social obligations. What do you think they are? Is voting one of them? How about going to college? Think about the connection between democracy and higher education and answer the following questions:

1. What do you think is the purpose of higher education?

2. Why did you decide to go to college?

3. Do you think your college education will help you become a better citizen? Why or why not?

4. Now go to the Campus Compact webpage at www.compact.org/resources-for-presidents/presidents-declaration-on-the-civic-responsibility-of-higher-education and read the "Presidents' Declaration on the Civic Responsibility of Higher Education."

5. Did your answer to Question 1 relate to the presidents' description of the purpose of public higher education? Why do you think it did or did not?

6. Why is an educated public necessary for a strong democratic society?

(Continued)

(Continued)

7. Is a public higher education attainable for all Americans? (Be sure to look at the information in Exercise 1.4 before you answer this question.) Why or why not? If not, what are the ramifications of this situation for our democracy? How can you, using what your sociological eye has uncovered, work to make public higher education more attainable and realizable for more people?

Exercise 1.2
Walking Billboards

You occupy many social roles in your life: You are a student, somebody's friend, somebody's child, and perhaps a parent, sibling, employee, team-mate, boss, neighbor, girlfriend or boyfriend, or mentor. To many thousands of companies out there, your main role is that of consumer. How do apparel companies market themselves, specifically, to men, women, transgender individuals, and those from different racial and ethnic groups? How do members of each of these groups act as "walking billboards" for the apparel companies that make the clothing they wear?

Next time you are in one of your other classes, note the following (Tip: try to choose a larger class to increase your sample size):

1. How many students are there?

2. How many are wearing visible product or company logos on their clothing, including footwear, baseball caps, and so on? (Include yourself in your answers.)

3. Are there any logos that occur more than once throughout the class?

4. Are there any logos or brands that are considered "in" on your campus?

5. Using your newly trained sociological eye, analyze the results you have gathered. What institutional and societal forces might be

at work here? Do you notice any specific trends along perceived racial or gender identities? Are there any apparent patterns in what different groups of people are wearing related to social messaging, teams, or cliques exhibited in their dress? How about the faculty? Do you detect any trends among your teachers? What does all of this tell you about consumerism, values, norms, and the culture of your campus?

Exercise 1.3
What's the Connection Between College Students and Sweatshops?

1. Watch *Outside the Lines: Cowboys Clothing Controversy* (at http://espn.go .com/espn/otl/story/_/id/7435424/dallas-cowboys-dip-sports-apparel- business-comes-allegations-sweatshop-labor or at https://youtube/ KP8kRqNP3ag).

2. Determine if your campus belongs to the FLA or the WRC. You can find out if your campus belongs to the FLA (http://www.fairlabor .org/affiliates/colleges-universities) and the WRC (https://www .workersrights.org/affiliate-schools/).

3. Ask the manager of your campus bookstore what vendors they use to obtain the clothes sold on campus. If your campus does belong to the FLA or WRC, contact that organization and ask about its findings on the vendors your campus bookstore uses.

4. If your school is not an FLA or WRC affiliate, do some research on your own to find information about the vendors. The USAS's Sweatfree Campus Campaign website, which you can access through http://usas .org/, is one useful site. You should also check out the websites for the Business & Human Rights Resource Centre (https://www.business- humanrights.org/en/find-companies) and Green America's Responsible Shopper (https://www.greenamerica.org/responsible-shopper).

5. Ask the campus bookstore manager if he or she is aware of the conditions under which the vendors' employees work.

(Continued)

(Continued)

6. Write a two- to three-page paper that describes your findings and why you believe your campus should or should not be affiliated with the FLA or WRC. If it is not (and you think it should be), describe which organization you think it should join and how you could help create a movement on campus to bring about such an affiliation.

7. Extra credit: conduct some research on the International Labour Organization's website at https://www.ilo.org/ and the Global March Against Child Labor's website at http://www.globalmarch.org/ to deepen your understanding of issues such as sweatshops and child labor. Expand your two- to three-page paper to a five-page paper, grounding it more deeply in the research, statistics, and reports you have uncovered.

Exercise 1.4
Worried About the Increasingly High Cost of Tuition?

You are not alone. According to the college board (Ma et al., 2017), after adjusting for inflation, from 2007 to 2008 through 2017 to 2018, in-state tuition and fees increased by $2,690 at public four-year colleges and universities and by $7,220 in private, nonprofit, four-year colleges and universities. Meanwhile, in 2015 through 2016, state funding to assist full-time college students was 11 percent lower (inflation-adjusted dollars) than in 2005 through 2006 and 13 percent lower than they were in 1995 through 1996 (Ma et al., 2017). Keep in mind that these figures do not include room and board costs.

1. How would strategies to deal with the increase in tuition vary depending on whether it is viewed as (a) a personal trouble or (b) a public issue?

2. Read at least four of the articles regarding student debt on the Inequality.org website at https://inequality.org/topics/student-debt/ to deepen your understanding of the issue. Expand your report,

grounding it more deeply in the research, statistics, and analysis you have uncovered.

3. Go to the website Higher Ed Not Debt (http://higherednotdebt .org). What are some of the actions suggested on that website or that you have thought of yourself that you could take to convince state legislators to increase funding for public higher education in your state?

4. Choose one of the actions you've listed under Question 3 that is a manageable action for you to take. Now, carry out the plan you devised and write a report that describes (a) what you did and (b) the outcome of your actions. Note that you might have to wait a while to complete (b), so you should start on (a) right away.

Exercise 1.5

Are Campus Workers Paid Enough to Be Able to Support Themselves?

Are workers on your campus paid enough to be self-sufficient?

1. Conduct research on the wages paid to cafeteria and custodial staff on your campus. You can gather this information through interviews, or if you are able to access the campus financial reports, you can use them.

2. Once you have gathered your data, read the Self-Sufficiency Standard at http://www.selfsufficiencystandard.org/ and look at the Self-Sufficiency wage data for your state. If your state is not listed there, you can use the Living Wage Calculator (http://livingwage .mit.edu/).

3. Based on what you have read, determine whether or not the cafeteria and custodial workers on your campus are making "living or self-sufficient" wages.

(Continued)

(Continued)

4. Read "How a Living Wage Is Calculated" (https://www
   .economist.com/the-economist-explains/2015/05/20/how-a-
   living-wage-is-calculated) and "Minimum Wage 101" (https://
   www.minimumwage.com/wp-content/uploads/2016/07/
   MinimumWage101_PolicyBrief_July.pdf) and look at the USAS
   overview of campaigns for Campus Worker Justice (http://usas.org/
   campaigns/campus-worker-justice/).

5. Write a two- to three-page paper that describes the findings for
   your campus, making sure to base it on your research, and how
   students can help raise the salaries of low-wage campus workers.

DISCUSSION QUESTIONS

1. Before reading this chapter, had you ever recognized a connection
   between your own life and the lives of people working in sweatshops?
   Why or why not? Do you now see a connection? Now that you have
   thought about this, how will you proceed?

2. Think of one personal trouble in your life (e.g., struggling to pay
   for college or to find a job that pays well, a time when you were
   discriminated against, etc.). Now, use your sociological imagination and
   relate it to a public issue. How would your attempts to deal with the
   problem differ depending on whether you view it as (a) an individual
   problem or (b) a societal issue?

3. Do you know what the president of the United States, your state senators, and
   your representative are doing about (a) sweatshops and (b) funding for higher
   education? If you don't know, why do you think you are unaware of their
   positions on these issues? Would their positions on these issues influence
   whether or not you would vote for them (or have they influenced your
   decision)? Why or why not? If you do know, what actions have you taken
   (if any) to try to influence them to adopt stronger positions and actions?

4. Do you think you could use your sociological eye to do something similar
   to what the members of the Social Justice League did at BSU? Why or why
   not? What are some of the ideas you have?

5. Explain why the second core commitment of sociology (social activism)
   must be preceded by the first (use of the sociological eye). Provide an
   example of how you might fulfill both core commitments.

SUGGESTIONS FOR SPECIFIC ACTIONS

1. Join an already established campus group working against sweatshops or working for the labor rights of workers on your campus. (You may substitute a different issue or campaign if you are already involved or interested in one, with the permission of your instructor.)

2. Establish your own campus group to fight sweatshops. Go to https://www.we.org/we-schools/program/campaigns/ and the anti-sweatshop movement's USAS website (http://usas.org) to learn how to form a local group to combat sweatshops.

3. Go to your representative's and senators' websites (you can find them at http://www.house.gov and www.senate.gov). Send them an e-mail or letter that conveys your thoughts and feelings about a public issue of concern to you.

4. Investigate the safety of the work and living environments provided for students, faculty, professional staff, and support staff on your campus. If there are clear deficiencies or inequities, organize a group of students, faculty, and professional and support staff to advocate for improved conditions for all campus workers. Sociology students at the University of Virginia, the University of California at Berkeley, and many other colleges across the nation have carried out just this type of research and activism. You can find information on recent campaigns at http://usas.org/ and in *Berkeley's Betrayal: Wages and Working Conditions at CAL* (http://publicsociology.berkeley.edu/publications/betrayal/betrayal.pdf).

   Please go to our website at http://study.sagepub.com/white6e to find further civic engagement opportunities, resources, peer-reviewed articles, and updated web links related to this chapter.

NOTES

1. If you did grow up in Sweden, Ethiopia, or Bangladesh, then imagine that you grew up in New Jersey or whichever state in which your university is located.
2. It works both ways, of course. And bonus points to you for looking at the endnote.
3. We should note that this view is less prevalent today than it was at the time of Sandra Day O'Connor's appointment to the Supreme Court.
4. To learn more about the history of the sweat-free campus movement read "The Future of the Student Anti-Sweatshop Movement: Providing Access to U.S. Courts for Garment Workers Worldwide" (http://digitalcommons.wcl.american.edu/cgi/viewcontent.cgi?article=1058&context=lelb). See also the WRC website (http://www.workersrights.org).

REFERENCES

Cable News Network. (2018, August 26). Same-sex marriage fast facts. Retrieved from https://www.cnn.com/2013/05/28/us/same-sex-marriage-fast-facts/index .html

Collins, R. (1998). The sociological eye and its blinders. *Contemporary Sociology*, 27(1), 2–7.

Egan, R. D. (2013). *Becoming sexual: A critical appraisal of the sexualization of girls* (p. 2). Cambridge, United Kingdom: Polity.

Fair Labor Association. (2018). *Affiliates: Colleges and universities*. Retrieved from http://www.fairlabor.org/affiliates/colleges-universities

Fair Trade Campaigns. (2018). Grow the fair trade movement in your community. Retrieved from https://fairtradecampaigns.org/

Friedan, B. (1963). *The feminine mystique*. New York, NY: W. W. Norton.

Gray, J. (2004). *Men are from Mars, women are from Venus: A practical guide for improving communication and getting what you want in your relationships*. New York, NY: HarperCollins.

Healy, V. O. (2012, August 29). Mature designs for young girls difficult to avoid. *Chicago Tribune*. Retrieved from http://www.chicagotribune.com/news/local/ct-met-mature-school-clothes-20120829,0,5946648.story

Hughes, E. C. (1971). *The sociological eye: Selected papers*. Chicago, IL: Aldine-Atherton.

Ma, J., Baum, S., Pender, M., & Welch, M. (2017). *Trends in college pricing 2017*. New York, NY: The College Board.

Mills, C. W. (2000). *The sociological imagination*. New York, NY: Oxford University Press. (Original work published 1959)

Park, M. (2016, August 31). 5 things to know about the Dakota Access Pipeline. Retrieved from https://www.cnn.com/2016/08/31/us/dakota-access-pipeline-explainer/

Pew Research Center. (2017, June 26). Changing attitudes on gay marriage. Retrieved from http://www.pewforum.org/fact-sheet/changing-attitudes-on-gay-marriage/

Rudick, D. (2004, May 18). Mass. marks first day of legalized same-sex marriage. Retrieved from https://www.bostonglobe.com/metro/2004/05/17/mass-marks-first-day-legalized-same-sex-marriage/lG0ZI0A2SG54q9Hww4lPDK/story.html

Schein, V. E. (2001). A global look at psychological barriers to women's progress in management. *Journal of Social Issues*, 57, 675–688.

Worker Rights Consortium. (2018). *WRC affiliated colleges and universities*. Retrieved from https://www.workersrights.org/affiliate-schools/

Founders and Foundations of Sociology
Theory

S ociology was founded by social scientists eager to (a) understand the major social changes of the late 19th and early 20th centuries and (b) make society better. In this chapter, you will learn how six of the founders of sociology—Karl Marx, Max Weber, Émile Durkheim, George Herbert Mead, Jane Addams, and W. E. B. Du Bois—carried out the two core commitments of sociology. Each of these theorists, in his or her own way, looked beneath the surface of society to understand how it operates and used this knowledge to improve society.

Although all of these founders responded to aspects of the social forces related to industrialization, their works are myriad and focus on a variety of subjects. Marx and Weber are considered conflict theorists,[1] Durkheim was a functionalist, Mead and Addams were symbolic interactionists, and Du Bois inspired critical race theory. However, to one degree or another, all of them looked at the roots of inequality in society and the possible solutions to this social problem. They used theories to explain how society works and how it might be improved. Like all explanations, some theories are more convincing than others. As you read through the chapter, think about which theories are most helpful to you as you try to understand how society operates and how you might work to make it better.

Karl Marx

According to Marx (1818–1883), class conflict over the control of the production of goods leads to inequality in society. He maintained that in every economic age, there is a dominant class (the owners) that owns and controls the means of production and exploits the other class (the workers).

For example, in the feudal era, there were landowners and serfs, and in the industrial era, there were factory owners and factory workers. Marx believed that the workers would eventually overthrow the owners when

1. the economic means of production was sufficiently technologically advanced that it could easily support everyone in society and

2. the workers united, realizing that they, as a class, were being exploited by the owners.

Marx believed that the workers (*proletariat*) were under a "false consciousness" regarding their social class arrangements. That is, although they were conscious that there were class differences, they didn't understand why these differences existed, how those in power had manipulated the system to create these differences, or even the extent of these differences. Thus, their *consciousness* of the class differences was *false*.

Marx believed that the owners (*bourgeoisie*) owned not only the means of production for market goods but also the means for the production of *ideas* in society. In Marx's words,

> The ideas of the ruling class are in every epoch the ruling ideas, i.e., the class which is the ruling material force of society is at the same time its ruling intellectual force. The class which has the means of material production at its disposal has control at the same time over the means of mental production, so that thereby, generally speaking, the ideas of those who lack the means of mental production are subject to it. (Marx & Engels, 1970, p. 64)

The owner's ability to control the dominant ideas of society helped them stay in power. The workers were exposed primarily to ideas that promoted the status quo and maintained their false consciousness. This worked to prevent the members of the proletariat from realizing that the capitalist system was designed to exploit rather than benefit them.

To counteract this false consciousness, Marx spent much of his life trying to unite the proletariat, encouraging them to establish a "class consciousness," overthrow the ownership society, and transform the economic system from capitalism to communism. Consciousness was key to Marx's approach. As long as millions of individual workers saw themselves as struggling alone or in competition with other workers, nothing would change. Marx wanted to impart a larger, societal view of the system to the working class, in which they would understand the role of the class system in their personal lives and act collectively against the system itself. His most famous attempt was *The Communist Manifesto* (Marx & Engels, 1848/2002), which concludes, "Let the ruling classes tremble at a Communist revolution. The proletarians have nothing to lose but their chains. They have a world to win. Workingmen of all countries, unite!" (p. 258).

Conflict theory is a modern extension of Marx's insights, although many conflict theorists support democracy, not communism. In its general form, *conflict theory* begins with the assumption that at any point in time, in any society, there will be different interest groups, different strata of society that have conflicting needs, and that much of what happens politically, socially, or economically is a manifestation of this conflict. Conflict theorists maintain that at the core of society lies the struggle for power among these competing groups.

Max Weber

Whereas Marx focused on class conflict and economic systems, Max Weber (1864–1920) looked more at the combination of economic and political power. Weber expanded Marx's idea of class into three dimensions of stratification: (1) *class* (based on possession of economic resources—most important in industrial capitalist societies), (2) *status* (prestige—most important in traditional societies), and (3) *party* (organizations formed to achieve a goal in a planned manner, such as political parties, unions, and professional associations—most important in advanced industrial, highly rational societies). In most eras, there would be a great deal of overlap among the three dimensions. For example, someone high in class would also tend to be high in status and political power.

Unlike Marx, Weber was very pessimistic about attempts to eliminate inequality from society. He believed that even if one aspect of conflict and inequality could be eliminated, others would remain and perhaps become an even more important basis for inequality (e.g., the rise of the importance of difference in party position in China after status inequality had largely been eliminated). Weber's (1946) definition of *power*—"the chance of a man or a number of men to realize their own will in a communal action even against the resistance of others who are participating in the action" (Weber, as quoted in Gerth & Mills, 1958, p. 180)—remains the starting point for most modern sociological explorations of power relations.

Weber's work on bureaucratic institutions helps us understand how power is won and held within advanced industrial societies of all types (whether capitalist, communist, or anything in between). Thanks to Weber, we now comprehend how powerful bureaucratic structures can be and how much of the structure remains intact even when the individuals in charge are replaced. For example, whoever controls the governmental bureaucracy in a highly developed nation can exert tremendous power over all aspects of that society. Controlling the government bureaucracy enables one to control key institutions in society (including the military) and to define the standards by which other bureaucracies will be created, evaluated, and carried forth.

The crucial element of a power structure is its perceived *legitimacy*. Persons or institutions have legitimacy when people accept their authority and follow their orders. Because the power of lower-level functionaries depends on the same system that empowers those in higher positions, it becomes difficult and dangerous and therefore unlikely for someone in a lower stratum to really challenge the upper strata. The bureaucracy protects itself.

When properly managed, bureaucratic structures are extremely efficient, whether they are being used well or poorly. One infamous example of this is that the same highly efficient train system that existed in

Germany before Hitler came to power (to transport workers and travelers) was used to transport men, women, and children to the death camps of the Third Reich. In fact, Hitler and the Nazis used many of the mechanisms of Germany's bureaucracy to carry out one of the most efficient (albeit horrible and nearly incomprehensible) acts in human history.

Although Weber cautioned against the establishment of sociology as a science that should direct society, he did not shy away from using his knowledge to try to guide his country (Germany) in turbulent times. He may have been pessimistic about anyone's potential to eradicate inequality, but he nonetheless felt obliged to do what he could to promote democracy in his society. Weber was deeply involved in the political realm throughout his life. His greatest impact on German society, as an engaged citizen, came toward the end of his life, during and right after World War I. He wrote many newspaper articles, memos to public officials, and papers against the annexationist policies of the German government during the war, and he advocated for a strong, democratically elected parliament and against the extreme ideologies of both the right and the left (Coser, 1977, p. 242).

Émile Durkheim

Whereas Karl Marx and Max Weber were conflict theorists, Émile Durkheim (1858–1917) adopted a functionalist perspective. According to this perspective, society is like a biological organism, with each organ dependent on the others for survival. Functionalists believe that society is made up of interdependent parts, each working for the good of the whole rather than being composed of competing interests (as conflict theorists maintain). Durkheim believed that humans are selfish by nature and must be channeled and controlled through proper socialization by institutions in society. According to Durkheim, properly functioning institutions, such as the education system, the family, occupational associations, and religion, will ensure that people work for the good of society rather than just for themselves as individuals.

Durkheim (1903/1933) maintained that society is held together by a sense of connectedness or *solidarity* that its members feel. This type of bond changes as society moves from simple (e.g., agrarian) to more complex (e.g., industrial and postindustrial). The simpler societies, in which almost everyone shares a common way of life, are based on what Durkheim called *mechanical solidarity*. In this type of society, there is little room for individualism. People are bound to one another through tradition and a common way of life.

The more complex societies, in which people perform different and often highly selective tasks, are founded on *organic solidarity*. In this type of society, people come together to exchange services with one another. It is the many exchanges and the interactions during those exchanges that bind the members of the group to one another. They rely on one another for needed goods and services and understand (to some degree) each other's different perspectives through communicating during their exchanges. In these societies, there is more room for individualism. However, while the members depend on one another to survive and prosper, the ties holding the community together are weaker.

Seeing the political and social upheaval that plagued his home country of France during his lifetime, Durkheim studied how society operates and sought ways to make improvements. According to Durkheim, at the core of a smooth-functioning society lies solidarity. Societies with increased divisions of labor can achieve stability only if their members are socialized through their institutions to believe that they are obligated to one another as members of a common community.

Durkheim argued that the existence of *external inequality* in an industrial society indicates that its institutions are not functioning properly. He divided inequalities into *internal* (based on people's natural abilities) and *external* (those forced on people). Because an organic society needs all its members to do what they do best in order for it to function most effectively, external inequality that prevents some people from fulfilling their innate talents damages all of society and should be eradicated. For example, if someone with the potential to find a cure for cancer—or just to be a good physician—never gets to fulfill this potential because she was raised in a poor neighborhood and attended a terrible school with teachers who never encouraged her to go to college, the whole society suffers.

Among Durkheim's concerns were the problems of how to reduce external inequalities and increase social consensus (solidarity). He maintained that it was up to the various institutions in society to create opportunities and incentives for all its members to become engaged citizens and share their gifts. Durkheim used his various positions in the educational system to mold France's public schools around these ideas.

George Herbert Mead

George Herbert Mead (1863–1931), the founder of symbolic interactionism, was the first sociologist to focus on how the mind and the self are created through social processes. Instead of looking at the individual as either distinct from or controlled by society, Mead saw that people are both shaped by and shapers of society. He was particularly interested in how the human self develops through communicating with others via language and other symbolic behavior (symbolic interaction).

According to Mead, humans are not truly human unless they interact with one another. In turn, the nature of our interactions with others determines how we see ourselves and our role in society. *Symbolic interactionists* maintain that society is a social construction, continually created and re-created by humans. We may not realize it, but society is maintained by our implicit agreement to interact with one another in certain ways. As we "practice" certain patterns of interaction, we reinforce the belief system that society "just works that way." Therefore, by changing how we interact with one another, we can change society.

Mead used his sociological expertise about the influence of the social environment to contribute to several social programs and movements in Chicago. For example, he served as treasurer of Hull House (the social service "settlement house" cofounded by Jane Addams and Helen Gates Starr; see below), was a member of the progressive City Club, participated in a variety of local movements and social programs in the city, and edited the journal *Elementary School Teacher*. Mead also spent much time advocating for a reform of the public school system that would provide tenure for teachers and give them greater influence over how they could teach students (Cook, 1993).

Mead spoke publicly and often on behalf of the immigrant population of Chicago, encouraging school reform to aid immigrants in the assimilation process.[2] He helped establish and, with Addams, served as vice president of the Immigrants' Protective League. The league supported studies on immigrants and pushed for legislation to protect them from exploitation. For many years, Mead was a fund-raiser and policymaker for the University of Chicago's settlement house and eventually served as president of the Settlement House Board of Directors. As a member of the board of directors, he led the effort to conduct research on social and economic conditions in the neighborhoods surrounding the stockyards in Chicago in order to advocate for social change. Mead's work to promote social scientific studies in Chicago eventually led to the creation of the Department of Public Welfare (Cook, 1993).

Jane Addams

Jane Addams (1860–1935) grew up in a time when the norm was for women to marry young and raise a family. Her father, however, a wealthy mill owner and state senator, permitted her to attend college, with the understanding that she would then marry and raise a family (Haberman, 1972). Instead, she ignored the gender norms of the time and courageously chose to become a public citizen and scholar. Addams is considered the founder of modern social work as well as one of the founders of sociology.

Addams, with her friend Ellen Gates Starr, established one of the first settlement houses in the United States. After visiting a similar neighborhood center in England during a trip after college, they returned to Chicago and created Hull House in 1888. Hull House became the model for "settlement houses" in newly developing urban areas throughout the United States. The model was based on the vision that middle- and upper-class people could move to the city and serve the poor while living among them. People in need, such as poor immigrants and women, would be exposed to the culture, values, and knowledge of the educated settlement house workers while they all resided in the same large households. Meanwhile, those who worked at the settlement houses provided social services and advocated for social policies that would empower and protect the members of these lower-class and working-class groups.

Hull House also served as "an underground university for women activists focusing on questions of housing, sanitation, and public health" (R. M. Berger, 1997, para. 5). Addams believed that all members of society, not only the privileged and wealthy, deserve protection. In turn, helping those in need benefits everyone in society, for "the good we secure for ourselves is precarious and uncertain until it is secured for all of us and incorporated into our common life" (Linn, 2000, p. 107).

One of the more colorful Hull House studies involved researching garbage collection in Chicago. Lack of proper collection was allowing disease to spread, particularly in poor, immigrant communities. In response, the Hull House Women's Club stepped out of the roles expected of them as ladies in late-19th-century Chicago and began to personally collect the garbage that was polluting the poor neighborhoods! Before doing that, however, they used their sociological eyes and research skills to carry out

> a major investigation into the city's garbage collection system. Then Addams submitted to the city government her own bid to collect garbage. The resulting public uproar forced the mayor to appoint Addams as garbage inspector for her ward. The Hull House women formed a garbage patrol, getting up at 6 a.m. to follow the garbage trucks, mapping routes and dump sites, and making citizens arrests of landlords whose properties were a health hazard. Their vigilance moved garbage reform to the top of Chicago's civic agenda, forcing industry to take responsibility for its trash. (R. M. Berger, 1997, para. 5)

This episode in the life of Hull House beautifully illustrates how Addams carried out the two core commitments of sociology. She and her colleagues used their sociological eyes to notice a pattern of inequality and then used social activism to address that inequity.

While Addams and her partners at Hull House helped initiate a grass-roots social reform effort, they soon recognized that structural changes were necessary to fix the causes and not just the symptoms of inequality. She and other Hull House residents worked to improve policy regarding the juvenile justice system, secure women the right to vote, strengthen workers' rights, and establish child labor laws. Addams and her colleagues knew that people without safe places to live and employment that paid a livable wage would never escape the cycle of poverty. While social services provided by the government assisted people in need, ending inequality would require structural changes such as greater access to education, skills training, and capital. Addams devoted her life to this effort. As a scholar and engaged citizen, she remained true to the belief that "nothing could be worse than the fear that one had given up too soon, and left one unexpended effort that might have saved the world" (as quoted in Lewis, 2009, para. 2). Addams was formally recognized as a leader in reforming social policy when, in 1931, she became the second woman, and the first from the United States, to receive the Nobel Peace Prize.

W. E. B. Du Bois

William Edward Burghardt Du Bois (1868–1963) grew up as one of the few black residents in Great Barrington, Massachusetts. While Du Bois was an excellent student, whom his teachers encouraged, his classmates did not treat him as an equal. During these early years in school, Du Bois began to recognize the racial division in society. He believed, though, that empowering himself with education would allow him to better understand and help improve the situation of black citizens. When Du Bois was in college at Fisk University, a historically black university in Nashville, Tennessee, he traveled with the glee club to a summer resort in Minnesota. There, he was exposed to privileged white vacationers and was struck by how their lives contrasted with those of the rural, poor black children whom he taught during the summers he was at Fisk.

Du Bois's early experiences furthered his drive to learn more and to improve the lives of black Americans in the United States. After leaving Fisk and earning a bachelor's degree at Harvard, he won a fellowship that allowed him to travel throughout Europe while he studied with the top social scientists of the time. In 1895, Du Bois became the first black person to earn a PhD from Harvard University. He went on to teach at a number of colleges and established the Department of Social Work at what is now Clark Atlanta University in Georgia.

Despite attending and achieving great success at Harvard, Du Bois (1953) described his experience as being "in Harvard but not of it." Being surrounded by but never connected to the mainstream academic and public worlds was a constant theme in the lives of both Addams and Du Bois. Even though he was armed with numerous scholarly achievements and a graduate degree from Harvard, Du Bois never gained a position as a full teaching professor at a predominantly white university. There is little doubt that Du Bois's and Addams's experience with discrimination shaped their desire to help those who suffered in their societies. They collaborated on numerous projects for social justice over the course of their careers (Deegan, 1988).

Throughout his life, Du Bois documented and railed against the status of black Americans, noting that although African Americans were an integral part of U.S. society, they were not fully accepted into it. Du Bois's major achievements as a sociologist, founder of critical race theory, and activist began with his famous study *The Philadelphia Negro* (1899). This groundbreaking work was the first social scientific research to dismiss the notion of racial inferiority. It documented the negative impact of racial discrimination and segregation on the condition of African Americans in the urban North.

Like Addams, Du Bois advocated for social policy changes and pushed for efforts to improve the situation of blacks throughout the country. He used the Hull House studies of poor immigrant neighborhoods in Chicago as his research model for *The Philadelphia Negro*. Also, just as Jane Addams and her colleagues at Hull House lived among the people they studied (a technique known as ethnography), Du Bois resided with the poor blacks in Philadelphia while he carried out his research for *The Philadelphia Negro*. While he did not enjoy living in that environment (Deegan, 1988), experiencing the conditions firsthand allowed him to establish rapport with and gain the trust of the thousands of black residents of Philadelphia he surveyed and interviewed.

Throughout his long career, Du Bois advocated for the rights of black Americans, women, and workers. He lived his belief that "there is in this world no such force as the force of a person determined to rise" (Aberjhani, 2003, p. 33). In addition to cofounding (with a number of people, including Addams) and leading the National Association for the Advancement of Colored People for many years, he promoted social action by editing the organization's journal, *The Crisis*; writing several books, including *The Souls of Black Folk* (1903/2005); organizing conferences of scholars; teaching; and speaking out on behalf of those without power in society. Eventually, he became frustrated with what he viewed as a lack of social progress in the United States. He joined the (then banned) Communist Party, surrendered his U.S. citizenship, and became a renowned citizen of Ghana before his death.

The founders of sociology were deeply interested in using their knowledge for the good of society. One wonders how they might make sense of society today and what changes they would recommend to curb current systems of inequality. As we can see, their respective views of the world influenced how they perceived the social issues of the day and their proposed solutions for them.

Theory and Society

Sociologists use theory to elucidate and make sense of social patterns. Without theories, we would have little or no understanding of why society operates the way it does and how we might improve it. Looking at the world through a theoretical perspective can also help us detect social patterns that we might otherwise overlook and help us figure out where we should concentrate our focus. For instance, conflict theorists are more likely to notice discrimination, class inequality, and struggles for power in society than are those who do not view society through a similar lens. Similarly, symbolic interactionists tend to be more aware of the impact of small-group interactions and symbols. For example, such theorists are quick to observe that seemingly minor behaviors (e.g., sitting with legs crossed or uncrossed) can have serious diplomatic repercussions.[3]

A subfield of symbolic interactionism is the sociology of knowledge, a theory that much of what we think of as "reality" in large part is seen that way because it has been "constructed" as what we *think we know* to be reality. For instance, one of the authors grew up Jewish and one grew up Catholic. In addition to this leading to us coming from families with differing ideas of God, each of our families also believed in differing "knowledges" on a variety of issues, from afterlife to gender equality to sexuality to how the earth and humans were created and so forth. Many of these are not only perceived knowledge of what is but also perceived knowledge of what is *right*, and thus what is wrong. Therefore, much of what we might learn from religion is not based on fact but on moral or cultural views, and yet these then contribute to what we often come to believe to be fact. Sociologists Peter Berger and Thomas Luckmann describe that "Everyday life presents itself as a reality interpreted by men and subjectively meaningful to them as a coherent world." Further, that "as sociologists we take this reality as the object of our analyses" (P. L. Berger & Luckmann, 1967). What is perceived as reality, thus happens inside the social world and is shaped and created inside this context. As sociologists, we are interested in studying the social forces that create our perceived knowledges and the outcomes these socially constructed knowledges then produce.

The level of analysis also varies among the theoretical perspectives. Some portray the world through a wide-angle lens, looking at larger (macro- and meso-) social patterns (e.g., functionalism and conflict theory), whereas others (e.g., symbolic interactionists) view society closeup, from a more detailed (micro-) angle. Sometimes, as the Sociologist in Action section below illustrates, theories help us achieve important practical goals. While theory is powerful in helping us analyze the social world, it is also important to remember that whenever people use a particular theoretical lens (or any point of view), they will be more attuned to some social patterns than others.

Sociologist in Action

Brian J. Reed

Social network theory helped the U.S. Army capture Saddam Hussein. Major Brian J. Reed, a doctoral candidate in the Department of Sociology at the University of Maryland, College Park, used his sociological training in social network analysis when he was stationed in Iraq and assigned the task of devising a strategy to capture the former Iraqi dictator. He described his use of social network theory in locating Hussein in the following way:

> The intelligence background and link diagrams that we built [to capture Hussein] were rooted in the concepts of network analysis. We constructed an elaborate product that traced the tribal and family linkages of Saddam Hussein, thereby allowing us to focus on certain individuals who may have had (or presently had) close ties to [him]. (Hougham, 2005, p. 3)

Reed's expertise in network analysis allowed him and the soldiers under his command to re-create and study a detailed picture of Hussein's social network, thereby, determining where he would be most likely to hide.

Reed also maintains that his sociological training helped him comprehend the Iraqi culture and, because of this understanding, more effectively carry out military operations in that country. Recognizing the practical applications of sociological research and theory, the Army Research Institute gave $1.1 million in 2005 to the University of Maryland's Center for Research on Military Organization, of which Reed is a member, to conduct research on social structure, social systems, and social networks.

Exercise 2.1
What Would the Founders of Sociology Say About . . . ?

1. Consider one of these issues of inequality today (the movement to privatize the world's water, the reduction of federal and state aid for state colleges and universities, the vast gap between races in wealth and income in the United States, the high cost of running for public office, the gap in wages between women and men in the United States or in the world, environmental racism, the increasing gap between the wealthy 1 percent and the remaining 99 percent, the increase in hunger in the United States, the differences in access to health care for those who are poor from those who are middle class, upper class, and wealthy, or the decrease in funding for housing for poor people).

2. Describe how three of the six founders of sociology discussed above might respond when told about this issue. Be sure that your answer also briefly summarizes what the issue is about.

3. Which of the responses makes the most sense to you? Why?

4. Is there anything you feel is missing from the founders' perspectives that you might want to add? If so, explain what you would like to add and why. If not, explain why you think their perspectives do not need to be expanded or revised.

Exercise 2.2
Going Deeper

1. Read the articles "Major US Immigration Laws, 1790-Present" at https://www.migrationpolicy.org/research/timeline-1790 (you'll need to download the fact sheet) and "Key Facts About US Immigration Policy and Proposed Changes" (http://www.pewresearch.org/fact-tank/2018/02/26/key-facts-about-u-s-immigration-policies-and-proposed-changes/).

2. Pick two of the theorists or schools of theory in this chapter. Write a two- to three-page paper outlining how each of these theorists or theoretical schools of thought would analyze both the history of and current debate surrounding immigration policy.

3. Which theorist or theory most helps to inform your own understanding and analysis of the immigration debate and why?

Exercise 2.3
Are You a Worker or an Owner?

1. According to Marx's perspective, do you think you were raised by workers (proletariat) or owners (bourgeoisie)? What makes you think so? Be specific.

2. Do you think most workers in the United States have developed a class consciousness? Why or why not? Be sure to provide evidence for your answer.

3. Can you see yourself encouraging workers to overcome their false consciousness and develop a class consciousness? Why or why not? How might you (or someone else) go about helping them do so?

Exercise 2.4
Just and Unjust Laws

1. Read Dr. Martin Luther King's "Letter From a Birmingham Jail" (http://okra.stanford.edu/transcription/document_images/undecided/630416-019.pdf).

2. Write a two-page paper discussing and analyzing this seminal piece, widely considered one of history's most important writings on social justice.

(Continued)

(Continued)

3. In your paper, also consider the social construction of knowledge. How do unjust laws come to be the law of the land? How are social norms and mores that at one point in history are seen as just come at another point in history to be seen as unjust? What does this tell you about the social construction of knowledge? The social construction of your own knowledge and morals?

Exercise 2.5
Different Perspectives Lead to Different News

This assignment will require you to watch the news for at least 1.5 hours a day for one week. It will also require you to have access to a wide range of stations. Watch CNN, BBC World News (now available in the United States on most cable networks and accessible online), and Fox News, each for half an hour a day for one week.

As you are doing so, make notes on (a) what stories they show on the news, (b) how they portray the news events (e.g., positively or negatively), and (c) how they compare with one another. Pay attention to which stories are addressed by all three and which stories are covered only on one of the news outlets. As you do so, complete the questions below. Pick one news story that all of the networks carry, and answer the following questions:

1. How much time does CNN, BBC World News, and Fox News each give the story?

2. Are the events at the center of the story portrayed positively or negatively (or both) by each of the three networks? How do the positive and negative portrayals differ from one another?

3. How would your knowledge of the news story be different if you watched just one of the news networks?

4. How would your perception of the news story (whether the story was important, negative or positive, etc.) be different if you watched just one of the news networks?

5. After the week is over, compare the different news stations' perspectives on the world. Were you able to clearly perceive three different perspectives? If so, how did they differ? Research and analyze in detail why it is that those differences exist. What does this tell you about the objectivity of news and news stories?

Extra credit: Go to the website for the organization Fairness & Accuracy in Reporting (http://www.fair.org/index.php). Look through the articles, and identify stories you think are important but were not covered by the news stations you analyzed in Questions 1 through 5. What is your explanation for why these stories were not included, and what does this tell you about mainstream news coverage?

Exercise 2.6
Different Perspectives in the Online Media

Find one conservative media source (e.g., WorldNetDaily at https://www.wnd.com/, Intellectual Conservative at www.intellectualconservative.com, or Free Republic (http://www.freerepublic.com/) and one left-leaning source (e.g., Salon at www.salon.com, *The Nation* online at www.thenation.com, ZNet at https://zcomm.org/znet/, or *Mother Jones* at http://www.motherjones.com).

Read the headline stories for each at the same time of day for five consecutive days. Then, answer the following questions in a two- to three-page paper:

1. How similar are they in their editorial approaches? That is, to what extent do the different papers make similar decisions about which stories are most important? If they are similar, why do you think this is so? If they differ, why do you think they are different?

2. How would your perspective on society differ based on which of the media sources you read on a regular basis?

3. Discuss the possible impacts of these types of media on U.S. society.

(Continued)

(Continued)

4. What are two or three major things your answers to Questions 1 through 3 tell you about print media today?

Extra credit: Choose any two of the theorists from this chapter (Marx, Weber, Durkheim, Mead, Du Bois, & Addams). How might each of them answer Question 3?

Exercise 2.7
The Media and Climate Change

Watch the following video and read the following overview piece about climate change: "Global Warming" at https://www.youtube.com/watch?v=ScX29WBJI3w#at%3D81 and "Global Climate Change" at https://climate.nasa.gov/evidence/ (read each section: Evidence, Causes, Effects, Scientific Consensus, Vital Signs, Questions, FAQ). Then, read "Competing Media Stories and US Public Opinion on Climate Change" (https://scholars.org/brief/competing-media-stories-and-us-public-opinion-climate-change). Then, read "What Are Donald Trump's Policies on Climate Change and Other Environmental Issues?" (https://friendsoftheearth.uk/climate-change/what-are-donald-trumps-policies-climate-change-and-other-environmental-issues).

Write a two-page paper that answers the following questions:

1. What are the key points you learned from the overviews and the article about climate change news coverage?

2. How has your own perception of climate change been influenced by the media?

3. Do you think reading this information and answering these questions have altered your perception of climate change? Why or why not?

4. Based on the article from Friends of the Earth, what is your analysis of the Trump administration's environmental policy in context of the other articles outlining climate change?

5. How might you use the information you have learned to teach students at your school about climate change? How, if you wanted to, might you use this to create action to create social change on this issue? Be specific.

DISCUSSION QUESTIONS ─────────────

1. Sociology has always been viewed a bit warily by leaders in most societies. Why do you think this might be? What is it about a sociological perspective that might feel threatening to those in power and those benefiting from the current system?

2. Imagine you are a sociological theorist. What social issue would you choose to study first? Why? Which of the three primary sociological perspectives (functionalism, conflict theory, or symbolic interactionism) do you think you would use to explain your findings? Why?

3. If, as symbolic interactionists maintain, society is merely a social construction (that is created and re-created anew through our interactions with one another), why is it so hard to address social issues effectively? How might a symbolic interactionist respond to this question?

4. Of the sociological perspectives covered in this chapter (conflict theory, functionalism, symbolic interactionism, and critical race theory), which do you think best explains inequality in the United States today? Why do you think so?

5. While Du Bois was able to attain a higher education only through scholarships, the other founders came from middle-class to upper-class backgrounds. Why do you think this might be? Do you think that most successful scholars today come from middle-class to wealthy backgrounds? Why or why not? If so, what are the potential repercussions?

6. Describe the difference between internal and external inequality. Provide an example of how external inequality can harm a society. What can (a) society and (b) you do to curb external inequality in your society?

7. Which of the theorists described in this chapter do you think best fulfilled the two core commitments of sociology? Why?

SUGGESTIONS FOR SPECIFIC ACTIONS

1. Many sociologists note that although sociological studies have pointed out good solutions to social issues they have been largely ignored by governmental leaders and the media. Go to the American Sociological Association website at www.asanet.org or the Society for the Study of Social Problems website at www.sssp1.org. Look around each site and find a study that provides a good basis for the use of sociological research in public policy.

2. Write a letter to your school newspaper or another local paper describing the study and what you think would be a good public policy based on it. Go to https://ctb.ku.edu/en/table-of-contents/advocacy/direct-action/letters-to-editor/main to get tips on how to write a letter to the editor.

3. Please go to the website for this book at http://study.sagepub.com/white6e to find further civic engagement opportunities, resources, peer-reviewed articles, and updated web links related to this chapter.

NOTES

1. Some consider Weber a functionalist.
2. See the University of Chicago Centennial Catalogue's faculty webpage on Mead at http://www.lib.uchicago.edu/e/spcl/centcat/fac/facch12_01.html.
3. In Arab nations, it is regarded as impolite to cross one's legs. In India, it is impolite to show the bottom of your shoe, as you would by crossing one foot over your knee instead of crossing your ankles.

REFERENCES

Aberjhani. (2003). *The wisdom of W. E. B. Du Bois*. Kensington, NY: Citadel.

Berger, P. L., & Luckmann, T. (1967). *The social construction of reality: A treatise in the sociology of knowledge*. Garden City, NY: Anchor.

Berger, R. M. (1997, July/August). The Good Housekeeping award: Women heroes of environmental activism. *Sojourners*. Retrieved from http://www.sojo.net/index.cfm?action=magazine.article&issue=soj9707&article=970722

Cook, G. A. (1993). *George Herbert Mead: The making of a social pragmatist*. Urbana, IL: University of Illinois Press.

Coser, L. A. (1977). *Masters of sociological thought: Ideas in historical and social context* (2nd ed.). New York, NY: Harcourt Brace Jovanovich.

Deegan, M. J. (1988). W. E. B. Du Bois and the women of Hull-House, 1895–1899. *American Sociologist, 19*(4), 301–311.

Du Bois, W. E. B. (1899). *The Philadelphia Negro*. New York, NY: Lippincott.

Du Bois, W. E. B. (1953). *Dr. W. E. B. Du Bois: Lecture on 23feb1953* (P. Goettlich, transcribed from a "Voices of Pacifica" recording). Retrieved http://www.mind fully .org/Reform/WEB-DuBois-23feb1953.htm

Du Bois, W. E. B. (2005). *The souls of black folk*. New York, NY: Pocket. (Original work published 1903)

Durkheim, É. (1933). *The division of labor in society* (G. Simpson, Trans.). New York, NY: Free Press. (Original work published 1903)

Gerth, H., & Mills, C. W. (Eds. & Trans.). (1958). *From Max Weber: Essays in sociology*. New York, NY: Oxford University Press.

Haberman, F. W. (Ed.). (1972). *Nobel lectures in peace 1926–1950*. Amsterdam, Netherlands: Elsevier.

Hougham, V. (2005, July/August). Sociological skills used in the capture of Saddam Hussein. *Footnotes, 33*(6), 3.

Lewis, J. J. (2009). *Jane Addams quotes* (Women's history—comprehensive women's history research guide). Retrieved from http://womenshistory.about.com/od/ quotes/a/jane_addams.htm

Linn, J. W. (2000). *Jane Addams*. Champaign, IL: University of Illinois Press.

Marx, K., & Engels, F. (1970). *The German ideology*. London, England: Lawrence & Wishart.

Marx, K., & Engels, F. (2002). *The Communist manifesto* (G. S. Jones, Ed.). New York, NY: Penguin Classics. (Original work published 1848)

Migration Policy Institute. (2013). *Major U.S. immigration laws, 1790–present*. Washington, DC: Author. Retrieved from https://www.migrationpolicy.org/ research/timeline-1790

How Do We Know What We Think We Know?

Sociological Methods

I n one memorable scene in the iconic 1975 comedy film *Monty Python and the Holy Grail*, a group of peasants drag a woman to the center of town to be burned at the stake. Sir Bedevere, the local lord, intervenes to ensure that justice is carried out.

Sir Bedevere:	What makes you think she's a witch?
Peasant 3:	Well, she turned me into a newt.
Sir Bedevere:	A newt?
Peasant 3:	[pause] . . . I got better.
Crowd:	[shouts] Burn her anyway![1]

How much is revealed in this simple scene!

First, this is the early Middle Ages, sometime around the mid-600s. This was a time of superstition, uncertainty, and fear for many people. It was a time before science, when religious spirituality contested with other spiritualities and forms of magic in explaining the world. Both Christians and pagans believed that witches walked the Earth and that one must be on guard against them. Certainly, there were no widely held assumptions about the nature of cause and effect or the logic of demonstrating that one thing caused another. Evidence did not need to be seen or examined; "proof" could come from the fact that a lot of people agreed on something. "Burn her anyway" is a strong argument when shouted by a large crowd of people.

The scene is carried by the self-reflective totality of its logic. Social scientists, in analyzing this social phenomenon, would start by acknowledging the importance of a belief system common during the Middle Ages that includes the existence of witches and humans being transformed by witches. We add an event: A man thinks he was turned into a newt, and those around him accuse someone of being a witch. Challenged to support the claim (*she* is a witch), they resort to the original definition of the event (there is a witch). Any evidence that worked in her favor could, of course, be interpreted to prove just how tricky those witches are. There is no space in the argument to consider counterclaims that she is not a witch or that there are no witches. You might as well light the fire. There is no way out of this logic.

Of course, things have changed a little in 14 centuries. Fewer people believe in witches,[2] most serious accusations are dealt with in the court-room, science has considerably more support than magic does, and it rests alongside religion as a widely accepted method of explaining the world. But scientific explanations rely on the logic of cause and effect, on the careful definition of our terms and assumptions, and on the idea that evidence can be tested. Even a casual observation of U.S. society suggests that these concepts are not as widely understood or applied as we might hope. One means of finding out if the research we read about is trustworthy is to check and see if the researchers followed the basic steps of the scientific process.

The basic steps of all social science research are as follows:

1. Choose a research topic.

2. Find out what other researchers have discovered about that topic.

3. Choose a methodology (how you will collect your data).

4. Collect and analyze your data.

5. Relate your findings to those of other researchers.

6. Present the findings for public review and critique.

The Three Questions

In addition to following the basic steps of scientific research, social scientists must establish the answers to three basic questions about their work before they publicize their findings: (a) How do you know? (b) So what? (Why is it important we know this?) (c) Now what? (What do we do with this new information?) Let us consider each of these questions, from the last to the first.

The "Now What?" Question

Research is the path toward answering questions. Before we can discover new findings, we first have to raise a question. Furthermore, we like to assume that both the question and the answer are important and can influence society in some way. When we find the answer to our question, we should do something with that knowledge.

It has been popular to picture Newton "discovering" gravity as an apple fell on his head. It would be absurd to suggest that up to that point (mid-1600s, about one thousand years past the time in which the *Monty Python* skit was set), no one had noticed that things fall. But it is interesting to wonder how the observation (that things fall) became a question (How and why do things fall?).

Something in the social, political, or scientific world had changed so that it became useful to answer these questions. Did things fall any differently after Newton had named the process? No. However, he discovered new information that powerfully influenced his and future societies when he answered the questions about why and how objects fall. His general theory of gravity changed the world (Eves, 1976).

Your work does not have to change the way people view reality and the cosmos to have meaning. In the late 1950s, Harold Garfinkel (1967) attempted to explain the personal situation of a woman who had been born with male primary sexual characteristics (a penis), who in puberty had developed female secondary sexual characteristics (breasts), and who had to make a choice—accompanied by surgery—to define her sexual identity. Such cases are extremely rare and mostly unseen by the majority of our society (see Sax, 2002). One might think that sociologists would not be concerned with a lone case. Yet to appreciate and explain the case of "Agnes," Garfinkel had to address sexual norms, the power of "belonging," sanctions against deviance, and the strategies by which those with "abnormal" conditions may try to "cover" them. Agnes's biophysical condition combined with a vast and complex social world of meaning to create a problem that she had to solve. It was not, strictly speaking, the ambiguity of whether she was "really" male or female that defined the problem but the social need for her to be unambiguous about it.

Since the time of Garfinkel's (1967) study, intersex persons have gained recognition and more acceptance. However, they still face the stigmatization that accompanies those who do not fit into the either/or characteristics most people still associate with sex categories. Garfinkel's research raises questions about (a) the social categories within which personal identities and sexual identities are constructed, (b) how norms become normal and what happens when you violate them, and (c) how we define and use notions of what is or is not "natural." Of course, once we have raised these questions, we have to start looking at everything that is considered natural and everything that is feared or hated because it is seen as unnatural, and we have to recognize how these categories are socially constructed (and can be reconstructed).

Very few people in the United States are burned at the stake anymore, but hate crimes and vigilante assaults still occur frequently. In fact, in 2017 hate crimes were up by 12.5 percent in America's ten largest cities, and this was attributable both to the interference of Russian operatives with social media and online news sources and an increased rhetoric and policy of "othering" in the new political landscape. Anti-Latino, anti-Muslim, and anti-gender/transgender hate crimes rose during this period, while hate crimes against African Americans, Gay, and Jewish people persisted, with these three continuing as the top overall targeted groups (Levin & Reitzen, 2018).

Sociological research can help us understand challenges like hate crimes. Fundamentally, Garfinkel's research was not about Agnes. It was about what the rest of us *think* of Agnes and many other people and the things that frighten or confuse us or don't fit into the norms we have constructed for our society—for instance, how do Americans think about race, sexual orientation, gender identities, and religious differences today, and how are these beliefs shaped by the people and institutions around us? Answers to the "Now what?" question often challenge us, both as individuals and as a society, to rethink our most comfortable and unquestioned assumptions (norms and values) about the world.

Answers to the "Now what?" question can also help us figure out how to improve society. For example, if we can show that the placement of environmental hazards tends to predominantly affect communities of color (e.g., Bullard, 2000; Ewall, 2013), we can then advocate for rules that prohibit environmental racism. Likewise, if we can reveal that all parents, no matter their level of education, can increase the chances of their children attaining a high level of education by having books in their home (Evans, Kelley, Sikora, & Treiman, 2010), we can try to make that information available to all new parents, create innovate programs to support them, and, in turn, affect positive social change.

The "So What?" Question

When we prepare our research, we have to have some idea of what new knowledge we are after and why it may be useful. The "So what?" question is like the "Now what?" question. But where the "Now what (do we do with these findings)?" question asks how we might make use of what we have learned from our research, the "So what?" question asks why sociologists need to research something in the first place. In other words, why does this matter in the world?

To answer the "So what?" question, we do not need to show that lives hang in the balance. We simply need to show that there is something to be learned about our society or culture (or a different society or culture) that is worth knowing. Even abstract knowledge is valuable in the quest to understand our lives, our world, and how they interact, since even seemingly inconsequential behavior adds up, collectively, to create the totality of human behavior.

Take jokes, for instance, which are a common part of our society and culture. More specifically, let's consider jokes that are about race, gender, religion, sexuality, ethnicity, a person's body shape or hair color, and so forth. What we find funny (that part of the joke that makes us laugh) can actually tell us a lot about racism, sexism, transphobism, homophobism, ethnocentrism, and so forth. Suppose you were to assign yourself a task tonight to write down all of the jokes you know about these different groups of people and then ask your friends to share all of their jokes with you to write down as well. If you were then to categorize your jokes (e.g., "jokes about African Americans," "jokes about people with blonde hair," "jokes about people who identify as gay," etc.) and then look through the jokes in each category, do you think you might find some common themes? Do you think that some of these themes might point to stereotypes, stigmas, biases, and prejudices? Of course they would, and in identifying these trends, you might now have some analysis to add in answering the question "It's just a joke, So What?" This is why sociologists are interested in behaviors that otherwise might to some seem relatively inconsequential, because they provide insight into our larger culture, beliefs, and morals.

The "How Do You Know?" Question

This question is the big one, and most of the rest of the chapter will address it. To put it simply, we know that our findings are well founded if we design scientific studies that are theoretically sound and driven by data. We ask questions, determine the kinds of data needed to answer them, figure out a plan for getting the data, and define a set of clear and consistent criteria for finding the answers in the data. Only then do we actually start collecting any data. We will explain each of these steps in further detail below.

All of us can do research, but it takes special skills to conduct *social scientific* studies. Research done by social scientists (and physical scientists) seeks to answer questions for which answers are not already known. Going to AccuWeather.com on the Internet to check the forecast for tomorrow is not the same thing as conducting research on the weather. Likewise, we could do some research on what movies we should see this weekend by checking out reviews at http://www.rottentomatoes.com. However, we will not have done social scientific research.[3] Social scientific research is a tool used to discover *new* findings about our social world.

Social scientific research is not carried out exactly like studies conducted in the natural or physical sciences.[4] We study things that are always changing and always dependent on the time and place in which we are looking. A physicist can ask, "What is the boiling point of water (at a given atmospheric pressure)?" A sociologist cannot ask, "What is the boiling-over point of social unrest?" There isn't any such thing that can be nailed down. Instead, we have to ask, "How does the occurrence of a single event of great injustice contribute to the likelihood that social unrest will lead to collective social action?" The answer is not a yes or a no or a number. It is a description of a social process in which some events lead to the possibility of other events. It's a probability estimate, but it is not a guess. We can show that our answers are reliable (consistent) and valid (accurate). Our questions are about how things work and why they work as they do. And the most interesting part is that no answer is ever final. There are always new circumstances, such as differences in history, culture, politics, economy, and geography, to take into account.

Once we have our questions, we have to figure out what data will answer them. If, for example, we wanted to know whether people in the United States generally want lower taxes, we could conduct a national *survey* with a couple of tax questions on it. You have probably seen surveys. Most are collections of short, simple questions on a set of related topics, in which you choose your answer to each question from a short list of choices. Surveys allow researchers to collect a few basic measures of something from a very large number of people, in a short amount of time, and in exactly the same way. (If you collect data from some people one way and from other people a different way, it is not appropriate to treat the data as equivalent.) The answers given by any individual *respondent* (the person who responds to the questions) are not really important to us. What we really look at is the pattern of answers across hundreds or thousands of cases. Surveys that contain questions followed by sets of answers from which respondents are asked to choose give us *quantitative* (or numerical) data on which we can perform statistical analyses. Statistical data give us the big picture of trends and social changes.

Perhaps we already have plenty of research that suggests that people want lower taxes. So now we might want to know what Americans are willing to give up for lower taxes (because lower taxes mean less governmental income and hence, usually, less governmental spending, meaning fewer programs). One might think that this could be done with a survey if we make the format more complex. We could, for example, identify long lists of things that the government spends money on and ask people which ones they would be willing to do without if they could have lower taxes. But then, how do we choose the list? And if we ask, "Which items would you trade?" how do we know that we aren't leading people to just identify the programs that they dislike, regardless of what they think about taxes? So just asking questions is not always the best way to find information.

And only asking people what they would be willing to cut excludes questions about what they would be willing to pay *more* for and thus would be biased. For instance, we know from some research that 84 percent of Americans are willing to pay increased taxes or tolls to fund roads and other transportation infrastructure if they know the taxes will go directly toward these items (CNBC, 2017). Similarly, 70 percent of Americans are willing to have their taxes raised by $177 a year to help reduce global warming, more than half are willing to pay more in taxes a year to support raises for teachers, 78 percent are willing to pay at least $200 more in taxes each year to support federal family leave programs, and of those who are insured, 75 percent would pay more in taxes for Medicare to cover dental care, 72 percent for Medicare to cover long-term care, and 77 percent for Medicare to cover eye care (Kotchen, Turk, & Leiserowitz, 2017; Feldman & Swanson, 2018; Ekins, 2018; NORC, 2018).

Often, as researchers, we need to know more than just whether people do or do not approve of something. We need to know *why* they do or do not approve. We need to sit down and listen to people. This process is called an *interview*. Interviews can be structured, semistructured, o unstructured. Structured interviews entail asking respondents a set list of questions, often with set answer choices. Unstructured interviews involve asking people a series of *open-ended* questions on the research topic and letting them answer in whatever way they like. Semistructured interviews consist of some set questions but also allow for the interviewer to add follow-up questions as needed, and they give respondents the opportunity to answer in nonstandard ways. *Probe questions* are often used in interviews to encourage the respondents to dig deeper in their answers and analysis and also to keep them from straying too far off the topic. The data gleaned from unstructured interviews are not a pattern of "yes" or "no" opinions but the respondents' own words. This type of data collection is *qualitative* and elicits data based on words that describe "qualities"—feelings, opinions, and firsthand accounts.

Unstructured and semistructured interviews allow us to listen to and question respondents at length and investigate intangibles, such as *why* they feel as they do. A survey might ask whether the respondent thinks that our nation should spend more or less than it does on loans to other nations. But we can't determine for sure *why* the individual believes that we should spend more or less by asking a survey question. An unstructured or semistructured interview gives more depth and more context than a survey, but it is longer, harder to carry out, and more expensive. Also, because time and money constraints usually mean that we can interview far fewer people in person than we could survey online or through some other means, it's difficult to make good generalizations from unstructured and semistructured interview data.

Surveys and interviews rely on *self-reported data*. This means that people tell us the answers. Yet research and common sense both tell us that people do not always know why they feel or act as they do or how they will respond to certain circumstances. Few of us want to think of ourselves as people who will walk away from someone in need of help or cause harm just because we have been so instructed. If you ask people what they would do under circumstances in which strangers need help or in which their own actions cause harm, what do you think they will tell you?[5] For this reason, sometimes you may need to observe without asking.[6]

Experiments are one technique for observing how people will react to different conditions. In experimental research, the researcher measures the relationship between two variables by manipulating one of them and observing the other. If we want to know whether people will help a stranger in need, for example, we can create a situation in which someone appears to be in need of help. That's the *cause* part of a cause-and-effect relationship (the first variable). Then, we can control circumstances so that a variety of people, one at a time, encounter the situation. Some of them will help, and others will not. That's the *effect* (the second variable). As much as possible, we would create a controlled environment in which almost nothing else can affect the outcome. For example, to control whether the *subjects* (the people being observed) were in a hurry or not, we might set up conditions in which they think they are either late or early for a meeting with me.

The classic model of an experiment, which you have probably seen acted out in movies, involves researchers (in white coats) in a lab who look through a one-way mirror while subjects respond to different *stimuli* (the *cause* part of cause and effect). *Fieldwork* provides another way of measuring how people respond to stimuli. Fieldwork is, in some respects, the opposite of experimental research. It requires the researcher to go out into the real world, where things are happening that he or she does not manipulate at all.

The place where you choose to do your experimental work is called your *field site*. Field sites might include a park on a nice day, a campaign office during the course of a political election, spring training with a baseball team, an office, a courtroom, a classroom, or any other place where the kinds of activities in which you are interested happen. As researchers, we cannot control, or really change at all, any of the many things that affect what people do. We can't create the stimulus. But, unlike in a laboratory setting, we can observe things as they actually occur in people's lives.

There are other techniques for data gathering and analysis. For example, researchers use *content analysis* when they look for patterns among cultural artifacts (e.g., television shows, commercials, newspaper articles, pictures in magazines, textbooks). *Focus groups*, or interviews conducted with a group of people, are another means of collecting data.

As you have just seen, we can use numbers to make connections between politics and the day-to-day realities of people who suffer from the effects of bad decisions. But it is ethnographic fieldwork that perhaps best embraces the lived experiences of those whose lives and circumstances we study. *Ethnography* refers to the study of cultural information, such as values and meanings, as opposed to data about structures, politics, and economics. Elliot Liebow's classic 1967 study of an American ghetto, *Tally's Corner*, for example, was praised, when it was first published, for the author's ability "to grasp the native's point of view" while emphasizing that "these 'natives' are other Americans; that their society is his society" (Rainwater, 1968, p. 431). Feminist sociologists have turned to ethnography[7] as a way of overcoming the cultural biases of research itself while conducting research on cultural biases (Ribbons & Edwards, 1997).

Objective Research

Recently, we saw a political survey sent out by a senator to all of her constituents. Most of the questions asked respondents to rate certain proposals on a 5-point scale that consisted of 1 = *strongly disagree*, 2 = *disagree*, 3 = *neutral*, 4 = *agree*, and 5 = *strongly agree*. (This is called a *Likert scale*, which is a common way to solicit opinions, including political opinions.) But it was the questions that were remarkable. Many were of the form "How do you feel about the federal government running up huge debts in order to . . . ?"[8] The problem should be obvious: The question is a *leading question* to which the only acceptable answer is disapproval of the federal government. Yet despite the poor methodology used, the senator will be completely within her rights to later state that some overwhelming percentage of her constituents disapprove of whatever the issue is. Although it is an all-too-common tactic in political and marketing research, creating the answers you want is not at all scientific. Sociologists do not use these tactics, as our goal is to study issues *objectively* and to then present these findings (whatever they might be, and even if *subjectively* we would have hoped for different results) to the public.

As social scientists, we therefore need to guard against this sort of data abuse in many ways, but two particularly apply. The first is that we have to be open about our methods. We have to describe in detail where and how we got our data, including making our questions public. The second significant protection against misleading data interpretations is to define and describe clear and consistent criteria for data evaluation before we even collect our data. These criteria have to follow from our research questions.

As social scientists, we also have to demonstrate the reliability and validity of our findings. We need to let our readers know, through a clear discussion of how we collected our data, just how sure we can be that our data accurately represent the population studied (*validity*) and whether or not we can expect that other researchers, using the same methods, would find similar results (*reliability*). And we have to be honest, analytical, and trustworthy in doing so! Although not every study has to be high in both reliability and validity (depending on the methodology used, sometimes that is impossible), we must clearly indicate just how valid and reliable our findings are. Unless we do so, those who read and rely on what we present will not be able to accurately judge the veracity of our findings.

Returning to the example of political policies, suppose we want to determine how popular or unpopular a group of policies is with voters. We would start by designing a set of procedures for measuring people's opinions. This might be in the form of a survey with a number of Likert-scale questions on it (though not the questions from the survey we have just seen). Recall that we had defined a 5-point scale, in which Answers 1 and 2 indicated disapproval, 3 was neutral, and 4 and 5 indicated approval. In this case, the analysis is easy. When all of the respondents' answers to the same question are averaged, the number can be interpreted on the same scale: Higher than 3 indicates approval. Of course, real studies often require much more complex criteria involving the interactions of many variables under different conditions. But the process always involves first laying out the criteria by which the data will be used to answer the question, then collecting the data, and only then drawing conclusions. When these steps have been taken, and the objective results uncover social problems, the second core commitment kicks in and sociologists can then use their knowledge to look toward solutions.

Sociologist in Action

Joan Mandle

Joan Mandle has used her sociological eye to engage students in feeling politically empowered inside our democracy. Her research focuses on antiwar, student, and antipoverty movements of the 1960s, and when her son was himself a college student, he separately became convinced that young people are the key to long-term social change. Her son, Adonal Foyle, at the time, was a star on his college basketball team, and later he went on to a 13-year career in the NBA.

Together, they formed a nonpartisan student organization Democracy Matters, with the goals of training young people to be effective citizens and getting big money out of politics to create fair elections. Foyle explains "We wanted to harness the collective power of young people, and make their voices heard politically so that they could become effective social change advocates for themselves and for others." The organization has engaged thousands of students on over 500 campuses, advocating against the powerful role that special interest groups play in campaign contributions and outcomes.

Mandle has served as executive director of the organization and has used her sociological research and knowledge of social movements to help build a national

(Continued)

(Continued)

movement of students on college campuses advocating for fair elections. The organization's work "nurtures a lifelong commitment to civic action and social change . . . (and) prepares students for political involvement and leadership in a wide range of civic activism—from campaign finance reform, to social justice, civil rights, workers rights, and environmental reform" (Democracy Matters, 2018).

Note: Quotes excerpted from Mandle, Joan. 2013. "Democracy Matters: Giving Students a Political Voice," in *Sociologists in Action: Sociology, Social Change, and Social Justice.* SAGE.

Exercise 3.1
Ethnography

1. Choose a public setting in which you can sit and take notes for at least an hour. The setting should be open to anyone and have lively social interaction. It must also be very unfamiliar to you, so that you can experience it as an "outsider." Begin your observations by writing out a general description of your setting. Who is there, and what are they doing? What sorts of activities go on in this place?

2. Describe how you fit into the setting. Are people noticing that you are observing them? How do you think these people view you?

3. For a 15-minute period, write down everything that you see happening. Try not to read anything into what you are observing. Observation notes might include things such as the following: "Two adults (1WM, 1WF [shorthand for 'one white male, one white female']) walk in pushing a stroller with a very young child in it. They talk. The man goes to the counter while the woman takes a table. The man orders and purchases two drinks and brings them to the table." Notes should not incorporate a lot of interpretation, the way this example does: "A couple with a child comes in. The husband asks his wife what she wants and then goes to get the drinks." You don't really know those things. You know that the two people are adults, but you

don't know their relationship or the details of their conversation. Don't write down assumptions about what you see—just what you actually observe.

4. After your notes are complete, try to make sense of them: (a) What patterns of interactions did you see? (b) What do you think they reveal? (c) Why do you think that? (d) What other explanations can you come up with? and (e) What would you need to observe or learn in order to know which explanation is the most likely?

5. What are the advantages and disadvantages of this type of data collection?

6. Would this be your preferred mode of data collection to understand what you observed? Why or why not?

7. Alternative assignment. Follow the above Steps, 1 through 6, but in Step 1, instead be very intentional to pick a place that is outside of your own normal space regarding race, sexuality, or religion. So, if you are white, perhaps attend an event put on by the Black Student Association; if you are Christian, perhaps choose a predominantly Muslim neighborhood; if you are heterosexual, perhaps choose a LGBTQ restaurant or social event.

Exercise 3.2
Gender and Leadership

1. Write a 10-question survey about the differences between those who identify as men and those who identify as women. Three of the questions should deal with issues of leadership and decision-making. For each question, offer respondents the same three answers to choose from: yes, no, or not sure. Make sure that your questions are worded in a way that is not leading.

2. Survey 10 people, and summarize the results. Describe what these results reveal.

(Continued)

(Continued)

3. Interview three people. Ask them the same questions about leadership, but for each answer they give, ask them why they think so. Do not lead them to answer in any particular way. Just ask them to explain themselves, and let them say whatever they want. Write down their answers in as much detail as you can.

4. For each of your interviews, analyze the reasons people give for their beliefs.

   a. How often do they express their answers in terms of personal beliefs or personal experiences?

   b. How often do they express their answers in terms of anecdotes—stories that they might have heard about people they do or do not know?

   c. How often do they refer to information they learned in one of their classes (if your respondents are students) or in news reports from reliable sources?

   d. How else do people explain their beliefs?

5. What do the answers to these four questions tell you about how most people in your sample form their beliefs about (a) gender and (b) how gender is related to leadership?

6. Compare your findings from your interviews and surveys. What kind of information did you get from each methodology? What benefits can be obtained by using both qualitative and quantitative data collection when you research a topic?

7. Read "Americans' Views of Women as Political Leaders Differ By Gender" at http://www.pewresearch.org/fact-tank/2016/05/19/americans-views-of-women-as-political-leaders-differ-by-gender/ (make sure not to read this until you have completed Steps 1–6 above). How does this research help to inform your own results and analysis?

8. Write a two-page paper outlining your results and your providing analysis.

Crimes that can be prosecuted by the federal government as hate crimes consist of criminal acts motivated by bias against members of a particular race, gender, religion, sexual orientation, or ethnicity, or those with a disability. Hate crimes may also be prosecuted at the state level, but what groups are protected under hate crime legislation varies among the states.

For this exercise, do the following:

1. Listen to the following three stories from National Public Radio: "Are Hate Crime Laws Necessary?" at http://www.npr.org/2012/04/10/150351860/are-hate-crime-laws-necessary, "DOJ Downplays Expectation for Hate Crimes Law" at https://www.npr.org/2012/05/02/151832687/justice-department-downplays-hate-crime-law-expectation and "Sikh Temple Shooting Stuns Congregation, Country" (http://www.npr.org/2012/08/06/158199106/sikh-temple-shooting-stuns-congregation-country).

2. Research your college's policy on hate crimes. (This information will likely be in your student handbook or may be accessed by using the search engine on your college's website and typing in the keywords "hate crime policy.")

3. Interview five students to find out (a) what they know about hate crimes and, specifically, if they know the definition of a hate crime; (b) if they think that hate crimes legislation is necessary (and why); and, if so, (c) what groups should be covered (and why).

   Your research will likely yield the most interesting results if some, but not all, of those you interview come from historical out-groups that have often been targeted by hate crimes (e.g., African American, Jewish, Muslim, LGBTQ, immigrant, Latino).

4. Write a two-page paper analyzing your interviews in the context of the National Public Radio stories.

Exercise 3.4
Social Media and Social Well-Being

Over recent years, there is growing research and expert analysis that the use of social media is contributing to negative effects on the mental, social, and physical well-being of those who use it, and particularly those who use it regularly.

1. Read "6 Ways Social Media Affects Our Mental Health" at https://www.forbes.com/sites/alicegwalton/2017/06/30/a-run-down-of-social-medias-effects-on-our-mental-health/#3e3945772e5a and "Is Social Media Bad for You: The Evidence and the Unknowns" (http://www.bbc.com/future/story/20180104-is-social-media-bad-for-you-the-evidence-and-the-unknowns).

2. Conduct a small survey with 15 students on your campus. In the survey, ask questions that find out (a) which social media sites they use; (b) how many hours they estimate that they spend on social media each day; (c) if they feel that social media has any connection to their well-being or lack of well-being.

3. Read "Nearly a Third of Young People Say Social Media Makes Them Feel Lonely" (https://www.sundaypost.com/fp/nearly-a-third-of-young-people-say-social-media-makes-them-feel-lonely).

4. Write a two-page paper outlining your results, analyzing it based on these readings, and offering some solutions for any issues you have identified.

Exercise 3.5
The Pros and Cons of Different Methodologies

In this chapter, we've described a variety of methodologies (means of collecting data), including classic experiments, fieldwork, interviews, and surveys. Each has its own set of strengths and weaknesses.

In a one- to two-page paper, do the following:

a. Briefly describe each of the four methodologies (classic experiments, fieldwork, interviews, and surveys).

b. Describe the strengths and weaknesses of each methodology discussed in the chapter.

c. Discuss how you might use each methodology to examine the issue of academic cheating on your campus. What would be the advantages and disadvantages of each of these means of studying this issue?

DISCUSSION QUESTIONS

1. Think of a research question that explores the reasons behind a norm on your campus that most people do not question (e.g., If "hooking up" instead of dating is the norm on your campus, how might you research how that type of normative behavior is supported? Or if commuter students tend to be disengaged from campus activities, what social forces support that norm?). Think of a second research question that explores an issue of race, sexuality, gender, or social class on your campus and that would allow you to study it.

2. Based on what you have read in this chapter, how will you determine if a politician is providing you with trustworthy or misleading information?

3. Think of something that you and members of your family have long believed to be true. How would you test this belief? If your research led you to change your mind, how could you convince others in your family to reconsider their own assumptions on the issue?

4. Which methodology covered in this chapter would you use to determine your classmates' attitudes toward global climate change? Why?

5. Sociologists expose social problems through good, social scientific research. What social issue would you like to bring further attention to through social scientific research? Why?

6. If you were to conduct a study to determine if your campus has adequate supports for (a) students of color, (b) students who are experiencing hunger, or (c) transgender and non-binary students, how would you collect your data? Why?

7. Which methodologies covered in this chapter would typically be most useful for someone researching a topic from the perspective of (a) a conflict theorist, (b) a functionalist, and (c) a symbolic interactionist?

SUGGESTIONS FOR SPECIFIC ACTIONS

1. Find a website for an organization that advocates for or protests against something. Most such organizations have websites these days, so you can start by choosing an issue about which people fight (e.g., a flat tax, economic inequality, genetically modified food, and police brutality); then, search out the groups that have positions on the issue. Select a website used by one of the groups. Identify two or three major claims that they offer on the website in support of their position. Do they present real evidence to back up these claims? How well do they document their sources? Try to verify the data on your own. Also, try to find contradictory information. Do you think they are being honest and factual or that they are not telling the full truth? What makes you think so? If not, create your own website (or alternate form of presentation, such as a fact sheet or PowerPoint presentation) that describes the same issue in a more balanced way.

2. Research a policy decision recently made by your college or university. Ascertain how the policy was designed. In particular, determine what research was undertaken to decide if the new policy was necessary. Was the research done in a social scientific manner? Based on the research undertaken, do you think the university policy was clearly needed and well formulated? Write a letter to the editor of your school paper that describes your examination of the policy and the research that culminated in it. Be sure to include your evaluation of the research in your letter.

3. Do the same as in Question 2, but specifically research a policy related to diversity, inclusion, and equity. When you have completed researching the policy, prepare a PowerPoint presentation on the issue that could be used to present to your class or to your university's administration. Make sure the presentation is based on research and evidence and puts forth proposed policy changes based on this as well.

   Please go to this book's website at http://study.sagepub.com/white6e to find further civic engagement opportunities, resources, peer-reviewed articles, and updated web links related to this chapter.

NOTES

1. This dialogue is found in the Internet Movie Database (http://www.imdb.com).
2. However, it's important to remember that the passage of time does not always result in more rational thought among populations. Just three centuries ago, "witches" were being put to death in Salem, Massachusetts!
3. Similarly, "researching" what is the best car to buy by reading consumer guides like *Consumer Reports* is not conducting true social scientific research.
4. Sociology, like psychology, anthropology, and political science, is a social science. Physics, chemistry, and biology are natural sciences in that they study the physical world or nature.
5. If you think you have an answer to that question, it's likely you are incorrect. Some people will exaggerate one way; others will exaggerate the other way. Many will guess, and almost none will really know. But there will always be a lot of different answers if you ask a lot of different people!
6. Please note that observational research, like all research dealing with human subjects, must be approved by your school's institutional review board. In almost all cases, you will need to obtain written approval from the people you plan to observe, and you must provide assurances that no psychological or physical harm will come to them as a result of your research. See the book's website for more information about the ethics of social scientific research.
7. Many feminist researchers view traditional, social scientific, and quantitative research methods as constructed by men and influenced by a male-dominated view of society. Ethnography and other types of qualitative research, in contrast to quantitative work, tend to provide a more complete picture of and a greater voice for the individuals being observed.
8. This was not an actual question. We're exaggerating just a little to make the point.

REFERENCES

Brown, J. (2018, January 5). Is social media bad for you? The evidence and the unknowns. Retrieved from http://www.bbc.com/future/story/20180104-is-social-media-bad-for-you-the-evidence-and-the-unknowns

Bullard, R. D. (2000). *Dumping in Dixie: Race, class, and environmental quality* (3rd ed.). Boulder, CO: Westview Press.

Cameron, L. (2018, January 3). Nearly a third of young people say social media makes them feel lonely. Retrieved from https://www.sundaypost.com/fp/nearly-a-third-of-young-people-say-social-media-makes-them-feel-lonely/

Cohn, D., & Livingston, G. (2016, May 19). Americans' views of women as political leaders differ by gender. Retrieved from http://www.pewresearch.org/fact-tank/2016/05/19/americans-views-of-women-as-political-leaders-differ-by-gender/

Consumer News and Business Channel. (2017, September 11). Most Americans would hike taxes, tolls to pay for roads, survey shows. Retrieved from https://www.cnbc.com/2017/09/11/most-americans-would-hike-taxes-tolls-to-pay-for-roads-survey.html

Democracy Matters. (2018). Mission and history. Retrieved from http://www.democracymatters.org/about-us/mission-and-history/

Ekins, E. (2018). Poll: 74% of Americans support federal paid leave program when costs not mentioned—60% oppose if they got smaller pay raises in the future. Retrieved from Cato Institute website at https://www.cato.org/survey-reports/cato-institute-2018-paid-leave-survey

Evans, M. D., Kelley, J., Sikora, J., & Treiman, D. J. (2010). Family scholarly culture and educational success: Evidence from 27 nations. *Research in Social Stratification and Mobility, 28*(2), 171–197.

Eves, H. (1976). *An introduction to the history of mathematics* (4th ed.). New York, NY: Holt, Rinehart & Winston.

Ewall, M. (2013). Legal tools for environmental equity vs. environmental justice. *Sustainable Development Law & Policy, 13*(1), 4–13.

Feldman, C., & Swanson, E. (2018). More than half of Americans support pay raises for teachers, poll finds. Retrieved from PBS NewsHour website at https://www.pbs.org/newshour/nation/more-than-half-of-americans-support-pay-raises-for-teachers-poll-finds

Garfinkel, H. (1967). *Studies in ethnomethodology*. Englewood Cliffs, NJ: Prentice Hall.

Kotchen, M., Turk, Z., & Leiserowitz, A. (2017, October 12). Americans willing to pay a carbon tax. Retrieved from Yale Program on Climate Change Communication at http://climatecommunication.yale.edu/publications/americans-willing-pay-carbon-tax/.kotchen

Levin, B., & Reitzen, J. D. (2018). *Report to the nation: Hate crimes rise in U.S. cities and counties in time of division & foreign interference*. Retrieved from California State University, San Bernardino, Center for the Study of Hate & Extremism website: https://csbs.csusb.edu/sites/csusb_csbs/files/2018 Hate Final Report 5-14.pdf

Liebow, E. (1967). *Tally's corner*. Boston, MA: Little, Brown.

NORC. (2018, March 26). *Americans' views of healthcare costs, coverage, and policy* [Issue brief]. Retrieved from http://www.norc.org/PDFs/WHI%20Healthcare20Costs%20Coverage%20and%20Policy/WHI%20Healthcare%20Costs%20Coverage%20and%20Policy%20Issue%20Brief.pdf

Rainwater, L. (1968). Review of the book *Tally's corner: A study of Negro street-corner men* by Elliot Liebow. *Social Forces, 46*(3), 431–432.

Ribbons, J., & Edwards, R. (1997). *Feminist dilemmas in qualitative research: Public knowledge and private lives*. Thousand Oaks, CA: Sage.

Sax, L. (2002). How common is intersex? A response to Anne Fausto-Sterling. *Journal of Sex Research, 39*, 174–179.

Walton, A. G. (2017, October 3). 6 ways social media affects our mental health. Retrieved from https://www.forbes.com/sites/alicegwalton/2017/06/30/a-run-down-of-social-medias-effects-on-our-mental-health

Creating Civic Engagement Versus Creating Apathy

Culture

Have you seen the *SpongeBob SquarePants* episode where SpongeBob goes to Sandy's house for the first time? If you have seen the show, you know that SpongeBob is an underwater creature (a sponge) and Sandy is a squirrel (from Texas) who lives in an air-filled, underwater structure. SpongeBob did not appreciate the fact that he relied on water to survive until he stepped into Sandy's air-filled home, where he couldn't breathe, became brittle, and started to disintegrate. Similarly, most people do not recognize how much they rely on their culture until they find themselves immersed in somebody else's.

Whether we are aware of it or not, we rely on culture for almost everything we do. All our decisions—what clothes to wear, what to eat (and how to eat it), with whom to live, and what to do with our lives—are made based on our culture. Similarly, the languages we speak, the religions we follow, the sports teams we root for, and our goals for what will make our lives "successful" are all heavily influenced by our culture, as too are the ways we think about race, sexuality, gender norms, politics, the sustainability of our planet, and most everything we think we "know."

Generally, sociologists divide the study of culture into the following three categories: (1) values—what members of society deem important, (2) norms—the rules and expectations for behavior that guide people as to how to act in society, and (3) artifacts (material culture)—the tangible objects that people from a particular culture create and use. If values are what we believe in, what we want, and how we think our lives should be, norms are the set of rules we have established that govern how we can achieve those things we value. The values and norms that a society establishes will then influence which artifacts members of that society will develop and how they will make and use them.

Values and norms are *social constructions*, constantly created and re-created through interaction. Values and norms are produced and reproduced by social processes, often without plan, discussion, or decision. We just get used to doing things in a certain way that has been taught to us and modeled through our culture. These values change over time, and they differ from society to society. Although there are some values that are shared by many societies (e.g., most capitalist societies value individualism, money, and material possessions), values are not universal. For example, when a very trim and physically fit person we know spent a year in a poor South Asian nation as a member of the Peace Corps, the villagers with whom she lived were constantly encouraging her to eat and gain some weight. The people there associated her thin figure with poverty, a condition to be pitied. Meanwhile, back in the United States, she often experienced just the opposite reaction from the people around her. Many people envied her physique and wished that they could have a similar one! When one of us traveled to Lesotho, a small country in southern Africa, he was using a cane because of an injury. In Lesotho, he discovered that people who walk with canes are the most respected members of the society, those who have earned the cane through their wisdom. There is only a limited number of canes (and other supplies) in these communities, so they are given first to the elderly, those viewed to be the wisest members of the community. On his return to the United States, he was quickly reminded that having a cane in our country is a sign of injury and weakness—a stark contrast to the admiration and respect it brought him while in Lesotho.

Like values, social norms vary over time and from society to society. For example, in the contemporary United States, expected behaviors for men and women continue to change as new forms of interaction arise. Only a decade ago, it was commonly suggested that only desperate people might use the Internet to find a date, but "singles bars" were popular. Today, arranging dates through Internet services is much more socially acceptable or "normal." In fact, in 2014 nearly 20 percent of brides met their significant other through online dating (Paul, 2014). Moreover, studies have shown that married couples who met online were more likely to stay together than those who met some other way (Cacioppo, Cacioppo, Gonzaga, Ogburn, & VanderWeele, 2013).

Where people go on their dates also varies from culture to culture. Visitors strolling at dusk through a park in Cuernavaca, Mexico, are likely to find *every* park bench filled with young couples. Most young people there lack the money to gain access to the private (or at least indoor) places where people typically go on dates in the United States. The authors were surprised to find in their travels to the Maasai Mara region of Kenya that, in some of the villages, uncles often arranged marriages between two young people, who never even had the opportunity to meet prior to their wedding ceremony. Girls and boys are also socialized differently in different cultures. For example, even in the United States, girls from one culture may be trained to hide their bodies whereas those from another are taught to exhibit their physiques. Their manner of dress indicates that they are members of different subcultures, meaning that they are not only a part of the larger U.S. society but also part of a smaller subgroup within it that has its own norms, values, and artifacts.

Subcultures

To varying degrees, subcultures exist within dominant cultures (the culture upheld by the group that has the most power in a society). A subculture is a group of people with cultural patterns (values, norms, and artifacts) that distinguish them from the dominant culture. College students, firefighters, stay-at-home parents, first-generation Jamaican Americans, environmental activists, Mormon Americans, paraplegic athletes, white antiracism activists, and working-class Americans are some of the many examples of subcultures in the United States. Chances are that *you* belong to one or more subcultures as well! Subcultures often arise around shared backgrounds. Italian Americans in an Italian enclave of a big city may form a subculture, whereas more "integrated" Italian Americans may not be a part of that group or may teeter between the subculture and the dominant culture. Subcultures also grow around shared activities. Surfers form one subculture, bikers another, and activists, perhaps, a third.

Yet sharing an activity is not enough to ensure the emergence of a subculture. There must be enough interaction for a common base of knowledge, habit, ritual, meaning, and values to develop. For example, college students, as a group, tend to share certain values, norms, and artifacts that differ from those of people not enrolled in college. They may value high grades instead of big paychecks (perhaps as a means to getting big paychecks in the future), tend not to sleep during the same times of day (or for the same number of hours) as most Americans, and research and write papers instead of handling or crafting material goods for purchase. Importantly, they also recognize each other as members of a common subculture.

Although they have some noticeably different ways of thinking and behaving, members of subcultures share the same *guiding* values, norms, and artifacts of the dominant culture. The things they do and think that distinguish them do not threaten or work against the dominant values, norms, and artifacts of their society. For example, your classmates on the football team and your classmates who are members of the campus environmental group may learn to think and behave in some different ways. Through doing so, each group creates its own subculture. However, the majority of both college football players and students involved in the campus environmental group follow the dominant norms and share the dominant values of the school and the larger society. Likewise, first-generation immigrant Jamaican Americans may enjoy different types of food from the majority of Americans and may tend to be more family oriented than Americans socialized only under the dominant American culture. However, both groups (albeit to varying extents) share and follow most of the primary values of U.S. culture (e.g., material success, progress, freedom, and individualism).

Members of subcultures both influence and are influenced by the dominant culture. For example, second-generation Jamaican Americans will tend to hold and follow dominant American values and norms more than their parents did. At the same time, the larger culture changes somewhat as immigrant groups interact with other members of society. The increasing variety of ethnic foods becoming common fare for many Americans is a good example of how norms (in this case, what most Americans eat) change as the demographic makeup of the United States changes.

Countercultures

Groups of people who have cultural attributes that *oppose* those of the dominant culture are known as members of a counterculture. Such groups can be peace loving, like the "flower children" of the 1960s, or willing to use violence, such as the Ku Klux Klan and the neo-Nazis. Members of countercultures often find it difficult to avoid violent confrontations (and some incite them), simply because they stand in direct opposition to some of the core values, norms, and artifacts of the dominant culture. The violence that erupted during the American Indian movement (led by the group called AIM) of the early 1970s provides a good example of how the opposition of dominant and countercultural values and norms can lead to violent clashes. AIM is a countercultural group with views based on Native American philosophies that directly challenge many of the dominant U.S. values and norms. For instance, AIM challenged values such as individual ownership of property and faith in the integrity of the U.S. justice and political systems. AIM's goals included a review of more than 300 treaties signed between American Indian nations and the U.S. government (which AIM leaders argued were broken by the U.S. government), autonomy for tribal governments, and a return to the traditional culture of the Native American nations. The group organized events throughout the United States, including the largest ever American Indian gathering in Washington, D.C., in 1972. Here, they attempted to present a twenty-point position paper to the Nixon Administration, through their Trail of Broken Treaties campaign. Unfortunately, the administration refused to receive the document.

Those who challenge dominant cultural rules, beliefs, and values can face dramatic and harsh reactions from those who adhere to the dominant culture. For example, when many of the dominant U.S. values and norms (individual ownership of property, faith in the U.S. justice and political systems) were directly challenged by AIM, the U.S. government responded to this threat with powerful efforts to stop AIM by jailing leaders and surrounding protest enclaves with heavy weaponry, which resulted in several shootouts. One member of AIM, the Native American political activist Leonard

Peltier, was convicted and imprisoned for murder more than 40 years ago. However, he is widely believed to be innocent of the crime for which he was convicted and thus is thought to be a political prisoner.[1] In 2012, members of the Taliban, an Islamic fundamentalist political movement, shot Malala Yousafzai, a Pakistani girl, in the head and left her for dead. The Taliban control areas of Pakistan where Malala and her family lived. They do not believe that girls should attend school, and they terrorized those who refused to obey their commands to stay at home. Malala was attacked for representing a countercultural group of girls advocating for their right to education. Malala fortunately survived and now stands as an even more powerful and well-known voice for girls' rights. She describes her decision to speak up, despite the danger she faced, this way: "I could either not speak and die, or speak and then die. I chose the second one" (Hamilton, 2014, para. 6). Incredibly, Malala survived the attack and has continued to speak out as an activist for girls' rights globally. For her courage, skill, and leadership, she is the recipient of many awards, including the 2013 International Children's Peace Prize and the 2014 Nobel Peace Prize, and in 2017, she was named the youngest ever United Nations Messenger of Peace (BBC, 2017).

Other countercultures include extreme religious organizations, cults, or other groups whose beliefs or goals place them in opposition to national laws. Countercultures often are more or less visible depending on the particular political climate at any historical moment. For instance, since the 2016 presidential elections, hate groups and hate crimes have increased significantly in the United States, reflecting values and norms from the current political landscape. 2017 witnessed a rise in U.S. hate groups to 954 groups, up from 917 groups in 2016 and 784 in 2014. This included a tripling of the number of anti-Muslim hate groups, from thirty-four in 2015 to 101 in 2016, the first year of Trump's presidency (Potok, 2017). Hate crimes are also on the rise, with a 17 percent increase between 2016 and 2017, and with 60 percent of these being motivated by race (FBI, 2018).

Such countercultural groups range from fundamentalist Mormons,[2] who believe in polygamy, to radical Islamic groups, such as Al Qaeda, and extreme right-wing Christian fundamentalists,[3] who use terrorism to promote their beliefs. All countercultural groups reject the dominant values, norms, and artifacts and seek out different ways of living. In some cases, they also try to impose their own culture against the will of the other members of society.

Recognizing Cultural Patterns

Now that we know what a culture consists of, we can begin to look for the patterns that help us recognize the dominant norms, values, and artifacts of different cultures, subcultures, and countercultures. Members of almost any group of people who gather together on a regular basis create their own *microculture*. Think of your own groups of friends. You probably have ways of speaking to one another, inside jokes, group norms and behaviors, and perhaps even modes of dress that set you apart, somewhat, from others in the immediate dominant culture in which you live. The same is true for the group of people who constitute your school's student body. Every campus has its own microculture(s). You have probably heard people talking about the "campus culture." What they are referring to are the values, norms, and artifacts most people on their campus share, which are often different from those found on other campuses, in the town where their campus is located, or in the larger society.

Every college or university has a reputation that is, in part, derived from the campus culture. For example, one of us received his undergraduate degree at a school known for its strong liberal arts education, having a sizeable Jewish student population, a cut-throat study climate among the student body, and its commitment to people joining together as a community to work toward social justice. As you can see, although we often speak of "the culture" of a community, it is clear that contradictory norms and values can exist within one culture.

The campus culture included valuing intellectual discussion, intramural sports, attending speakers and events on campus, competition, academic achievement, and friendship; working for social justice; and partying in small groups in other students' rooms, as opposed to big, campus-wide parties. Some of the dominant norms were studying, playing sports, volunteering to help in the community, hanging out with friends, hooking up (rather than dating), drinking lots of beer, and wearing either athletic or alternative clothing. The artifacts of the campus included architecturally interesting buildings; a well-respected Department of Community Service; two libraries seemingly always crowded with students; cramped on-campus housing; run-down off-campus apartments nearby to party in; and an old European castle-turned-dorm with a popular coffee house. These artifacts are more than just objects; they are manifestations of the values (some contradictory) of the college and the collective membership of the college community. The architecture, classroom sizes, coffee house, and general layout of the campus, for example, all promoted the intentional value of community gatherings and creating space for intellectual conversations, including about social justice issues and thinking about social change.

Civic Engagement on Campus

Civic engagement appears to be on the rise on college campuses. Two thirds of all Americans between the ages of 18 and 24 used a social networking site (SNS) for political activity during 2014. The 2016 election saw a strong youth turnout, and these young voters made a major impact in the ballot box. Fifty percent of young adults between the ages of 18 and 29 voted, with 13 million favoring Hilary Clinton and 9 million voting for Donald Trump (THE CENTER FOR INFORMATION & RESEARCH ON CIVIC LEARNING AND ENGAGEMENT [CIRCLE], 2016). Do you know eligible young adults (or are you a young voter) who didn't cast their vote in 2016? Since young voters now account for 31 percent of the electorate and since young voters favored Hilary Clinton, a higher turnout rate would have likely changed the result, and we would have had a different president of the United States.

Today, evidence of student interest in worker justice issues can also be found on campuses all over the United States. In 2008, the University of Wisconsin Oshkosh was the first institution of higher education certified as fair trade. By 2018, 112 colleges and universities in the United States sold certified fair trade products in their dining rooms and campus shops (Fair Trade Campaigns, 2018). As of 2018, there were 193 colleges and universities affiliated with the Students Against Sweatshops–sponsored Worker Rights Consortium, an organization that monitors the enforcement of the codes of conduct signed by colleges and universities to ensure that clothing with their schools' logos is made in sweat-free factories. As discussed in Chapter 1, students at many schools (e.g., Wayne State University, Fairfield University, the University of Pennsylvania, the State Universities of New York, the University of Arizona, Boston College, Duke University, and California State College, San Bernardino) have convinced their university administrators to stop buying apparel made in sweatshops and to have the Worker Rights Consortium monitor the operation of the apparel factories they use (to prevent abuses of workers).

The Boot the Bell campaign, led by the Coalition of Immokalee Workers and supported by students across the nation, demanded that Taco Bell and its parent company, Yum, stop the practice of indentured servitude and increase the wages of tomato pickers. By the time it ended in victory in 2005, it had spread to more than 300 college and university campuses and 50 high schools (Student/Farmworker Alliance, n.d.). Since Yum Foods signed on in 2005, the following food companies have also pledged to participate in the Coalition of Immokalee Workers–initiated Fair Food Program and make sure the companies that provide them with tomatoes pay field workers an additional penny a pound and treat them fairly: Yum Brands (2005), McDonald's (2007), Burger King (2008), Subway (2008), Whole Foods Market (2008), Bon Appetit Management Company (2009), Compass Group (2009), Aramark (2010), Sodexo (2010), Trader Joe's (2012), Chipotle (2012), Walmart (2014), Fresh Market (2015), and Ahold USA (2015) (Greenhouse, 2014).

Students have also pushed colleges and universities to do their part to address global climate change. Among the college students surveyed by the *Princeton Review* (2018), well over half (63 percent) noted that "having information about a college's commitment to the environment would impact their decision to apply to or attend a school." Noting students' interest in this topic, the *Princeton Review* now publishes a guide on green colleges and provides similar information on its website, including an honor role of the "greenest" colleges and universities. If you attend College of the Atlantic, State University of New York (College of Environmental Science and Forestry), or University of Vermont, you will be proud to know that your school ranks as one of the top three greenest colleges.

Many students across the country also made their campuses *divest* (withdraw all investments) from any corporations that were doing business with the genocidal regime in Darfur, Sudan, during the first decade of this century. As of 2016, there were more than 850 middle school, high school, and college chapters of the Students Taking Action Now: Darfur organization, which organized students to stand up against genocide in Sudan, Myanmar (former Burma), the Democratic Republic of Congo, and throughout the world. Ultimately, this divestment movement played a significant role in ending, albeit tenuously, the genocide in Darfur.

University administrative leaders also realize the importance of teaching students how to fulfill their obligations as citizens effectively. Today, 250 state colleges and universities participate in the American Democracy Project, which focuses on providing service-learning opportunities for students. In addition, almost 1,000 college and university presidents have signed the Campus Compact pledge to "educate citizens." More than six hundred U.S. institutions of higher education have signed on to the American College & University Presidents' Climate Commitment to reduce global warming and to "green" their campuses. Sociology students can use the tools of the discipline to be at the forefront of these efforts and to make sure they are carried out in effective and just ways.

Social Networking Sites and Civic Engagement

In recent years, civic engagement among the general public has also been spurred on by SNS. A recent Pew Research Center (Anderson, Toor, Rainie, & Smith, 2018) survey indicates that the percentage of Americans engaged in political activity on SNS was 53 percent in 2018. One third of SNS users, or 32 percent of all adults, encourage others to take action on issues they care about; 18 percent changed their profile picture to an issue they care about; and 20 percent looked up information on local protests or rallies. Furthermore, half of all African Americans indicate that using SNS for political purposes is important to them and 69 percent of Americans believe that SNS is at least somewhat important for getting elected officials to pay attention to issues. Moreover, political activity online tends to lead to more political activity offline. For example, 43 percent of SNS users indicated that they were inspired to learn more about political topics and 18 percent reported that they were inspired to take action on a political topic because of what they had read online (Anderson et al., 2018). One example of the use of social media for activism is the #BlackLivesMatter campaign, for which the hashtag has been used over 30 million times, generally spiking in use when there are high-profile cases involving shootings of black men by police. Another is the #MeToo movement, giving voice to women who are survivors of sexual violence, with the hashtag having been used over 19 million times in late 2017 and throughout 2018 (Brown, 2018).

While some people have disparaged political activity online as "slacktivism," for its supposed lack of relation to offline political activism, the Pew study (Anderson et al., 2018) shows that this criticism cannot be supported. Online political activity has led many people to become more civically engaged. Forty-three percent of SNS users say that they decided to learn more about a political issue because of what they read on an SNS. Almost all those who are politically active online (83 percent) are active both on and offline (in the "real" world).

A Growing Ideological Divide Between Republicans and Democrats

Those who pay attention to politics and are engaged in the political process tend to be very ideologically divided (Abramowitz, 2010; Kohut, Doherty, Dimmock, & Keeter, 2011). The difference in values between Republicans and Democrats increased tremendously during the George W. Bush and Barack Obama presidencies and seem to have moved from a divide to a chasm during the Trump presidency. The divide between Democrats and Republicans is now greater than those across gender, race, age, and class. Among Republicans, there are twice as many conservatives as moderates. While liberals do not dominate the Democratic Party (as conservatives dominate the Republican Party), the percentage of liberals in the party has grown and now equals that of moderates (Kohut, Doherty, Dimmock, & Keeter, 2012).

The greatest divide between Democrats and Republicans lies in how the members of each party view the role of government; and this divide has increased dramatically over the past quarter century. For example, Republican support for the idea that government should aid the poor has dropped substantially over the past 25 years. In 1987, 79 percent of Democrats and 62 percent of Republicans believed that the "government should take care of people who can't take care of themselves," while in 2012, 75 percent of Democrats and just 40 percent of Republicans agreed with this statement (Kohut et al., 2012). These differing beliefs about government have contributed to a growing cultural divide between the two major political parties in the United States.

Sociologist in Action

Amy Lubitow

Amy Lubitow uses her sociological toolkit to reveal connections between environmental exposures and breast cancer. When she was completing her PhD in Sociology, she began volunteering with the Campaign for Safe Cosmetics, as she appreciated the ways that this organization and other nonprofits advanced clear, accessible information about the health effects of toxic exposures from food, water, and personal care products. She began a writing partnership with Mia Davis, an activist who served as the National Grassroots Organizer for the Campaign for Safe Cosmetics.

Lubitow and Davis were concerned with a phenomenon called "pink washing," where corporations place pink ribbons on their products as fundraisers and to illustrate their "corporate social responsibility," when in fact those very companies and sometimes the very products bearing the ribbon contain toxic chemicals linked to breast cancer. In effect, these companies co-opt this breast cancer symbol and distract consumers, drawing their attention to the "cure" and away from questioning the "causes" of breast cancer, the latter which could bring their products under scrutiny and decrease corporate profits.

Through their scholar-activist partnerships, Lubitow and Davis wrote and published their article, "Pastel Injustice: The Corporate Use of Pinkwashing for Profit," in the journal *Environmental Justice*. Their article exposed the mechanisms of this corporate practice and illustrated how it serves to reframe breast cancer as a health problem to be cured rather than a health problem that could be prevented through decreasing everyday toxic exposures. As is typical of great activist scholarship, Lubitow and Davis embraced the momentum of this publication to spread the word and advance their recommendations for structural change, including regulatory reform. They spoke at academic conferences, public venues, on the radio, and as guest bloggers on prominent health organization websites. Their findings were also reprinted in several popular print media venues, including Science Daily and Forbes.com. The article on Forbes.com even compelled a written response from

Avon, a major player in pinkwashing, which led to mobilization to hold the corporation more accountable. Together, Lubitow and Davis were able to combine their unique skill sets in sociological theory and research and activist strategic mobilization to expose a damaging practice and call for social change.

Sources

Lubitow, A., & Davis, M. (2011). Pastel injustice: The corporate use of pinkwashing for profit. *Environmental Justice, 4*(2), 139–144.

Lubitow, A. (2015). Breast cancer activism: Learning to write collaboratively for social change. In *Sociologists in action on inequalities: Race, class, gender and sexuality*. Thousand Oaks, CA: Sage.

Exercise 4.1
Student Survival Guide

Pretend that one of your classmates is new to the United States (and from a very different culture).

1. Make a list of what he or she will need to know about the dominant culture in the United States. Make sure to include explanations about clothing, food, music, television, and everything that a non-American might need to know in order to "make it" in the United States.

2. Once your list is complete, make a list of the norms and values that the items on your list represent.

3. Look at the list of norms and values and write a one-page essay analyzing what they tell us about U.S. society.

4. Now, repeat this exercise but pretend that you are writing it for a transfer student or an incoming first-year student to your campus who was born in the United States but who will need to adjust to the subculture on your campus.

Campus Culture: Seeing What's Around Us

In an approximately three-page paper, answer the following questions:

1. Think about your own campus. List five of the dominant values on your campus. Now list five dominant norms and five artifacts that reflect those values or contradict them.

2. Do you think your campus culture is different from or typical of most college campus cultures? How do you know this, or why don't you know enough to answer this question?

3. Identify some of the visible subcultures on campus. What makes each a subculture rather than just a bunch of people who have something in common?

4. Do you feel as though you are part of the dominant culture on campus? A microculture? A subculture? A counterculture? Some of these? All of these? Explain.

5. Why is a campus culture important? How does it influence how you spend your time and energy in college? How does it influence people of different races, gender identities, ethnicities, nationalities, sexualities, political beliefs, and so forth?

6. Do you think your campus culture has encouraged you to become a more active citizen? Are you engaged with civic activities on campus? Off campus? Why or why not? If so, is there something in your campus culture that has encouraged your active citizenship? Are there ways that this can be improved on campus? If not, is there something in your campus culture that has discouraged your active citizenship? If there is, what is it, and what do you think might help change this type of culture?

Exercise 4.3

A Consumer Culture?

1. Write down an extensive list of what material goals you hope to achieve by the age of 30 (a car or cars, house, wardrobe, lifestyle, etc.). Or,

if you are an older student, what do you hope to achieve by the next relevant decade?

2. Watch the video *What Would Jesus Buy?* (http://topdocumentaryfilms.com/what-would-jesus-buy/) and *The Story of Stuff* (https://storyofstuff.org/movies/story-of-stuff/).

3. Outline the major points made in each of the videos.

4. Now, analyze your own consumerist goals in light of the arguments made in the videos. Why do you think you have the consumerist goals that you have? What in U.S. culture drives you in your quest for material success? Are there any downsides to having these consumerist goals? What do cultures that emphasize consumerism less than ours have that you find to be enviable, if anything?

Extra credit: If you have concluded that Americans are too steeped in a consumerist culture, create a campaign to educate your campus about the issue. Produce a fact sheet to help argue your case with information you can find at http://www.newdream.org, http://www.adbusters.org, and the Small Planet Institute at http://smallplanet.org/, or from other readily available sources on the Internet.

Exercise 4.4
Political Culture Data Collection for Dorm Dwellers

Use participant observation to examine the political culture in your dorm.

1. Look at your dorm's list of organized events (social events, lectures, etc.) as well as unorganized, spontaneous events and attend as many of them over the course of one month as possible.

2. Try to participate in as many informal discussions as possible with your dorm mates during the same month.

(Continued)

(Continued)

3. At each gathering, carefully note the following: (a) what type of gathering it is (friends studying together, watching a movie, gossiping about relationships) or, if it's a dorm event, the topic of the event, and so on; (b) how many times topics regarding racism, sexism, heterosexism, religionism, and so forth, were raised (if at all); (c) how many times these topics could have fit into the conversation but were not raised; (d) the reactions of others if and when these topics were raised (did the person who raised them receive positive or negative feedback or sanctions?); (e) (if these topics were raised) whether someone mentioned participating in an action to address the issues being brought forth; and (f) the reaction to that suggestion.

4. What do your findings indicate about the political culture in your dorm?

Extra credit: Devise a plan to make your dorm a place that is more welcoming—and even encouraging—of political discussion. Be thoughtful, strategic, and specific.

Exercise 4.5
Participant Observation of a Political Protest or Other Political Event

Conduct a participant observation of a political event on campus.

1. Find out what student activist groups (officially recognized or ad hoc) exist on campus by talking to other students, professors, people in the campus activities office, and so forth.

2. Choose an organization whose goals you feel comfortable supporting. Do not "infiltrate" a group by misrepresenting your own views.

3. Participate in organizing and carrying out at least one campus event (protests count).

4. Throughout your time with the group, take note of how political issues are discussed. Do they represent dominant culture, subculture, or counterculture values and norms? Elaborate.

5. Which conversational topics are encouraged and which are discouraged (result in sanctions) in the group?

6. What are the reactions to the efforts of the group from fellow students, professors, professional staff, and administrators?

7. What do your answers to the questions above tell you about the norms and values of (a) the culture of the group in which you participated and (b) your campus culture?

Exercise 4.6
Changing Body Expectations of Women Throughout History

Body expectations and body image are a reflection of cultural values and norms and have undergone consistent changes throughout history.

1. Read the article "The History of the 'Ideal Woman' and Where That Has Left Us" (https://www.cnn.com/2018/03/07/health/body-image-history-of-beauty-explainer-intl/index.html).

2. Write a one-page summary of what you have learned from this article, including your own analysis. What did you not know? What did you find most interesting? Did the article change any of your thinking?

3. Next, read the "Key Findings" section (pages 5–8) in the research report "Children, Teens, Media and Body Image" at https://www.commonsensemedia.org/research/children-teens-media-and-body-image (click on "Download the Full Report" toward the top).

4. Write a one-page paper summarizing and analyzing the key findings of this study.

Exercise 4.7
Globalization: The Great Cultural Divider?

Go to David Brooks's column "All Cultures Are Not Equal" (2005) at http://www.nytimes.com/2005/08/10/opinion/11brooks.done.html?_r=0 and answer the following questions:

1. Do you agree with Brooks that U.S. citizens are increasingly separating into distinct and isolated subcultures? Do you think that this trend could lead (or has led) to more countercultures in the United States?

2. Write a one-page outline of how you would carry out research to test one (or more) of Brooks's statements in the above article.

Exercise 4.8
Hate Crimes and What They Reveal About Our Culture

1. Read the three articles "Two Decades After Matthew Shepard's Death, 20 States Still Don't Consider Attacks on LGBTQ People as Hate Crimes" (https://www.cnn.com/2018/10/12/health/matthew-shepard-hate-crimes-lgbtq-trnd/index.html) and "Is It Safe to Be Jewish in New York?" (https://www.nytimes.com/2018/10/31/nyregion/jewish-bias-safety-nyc.html) and "Violence Against the Transgender Community in 2018" (https://www.hrc.org/resources/violence-against-the-transgender-community-in-2018).

2. Write a one-page summary of the three articles.

3. Then, also write a full one-page analysis of the three articles. What do each of these articles tell you about culture, values, subcultures, and countercultures in the United States? What solutions do you propose to changing the values you have identified and in moving toward a more inclusive and equitable society?

DISCUSSION QUESTIONS ────────────────

1. Describe a subculture to which you belong. What norms, values, and artifacts distinguish it from the dominant culture? How does belonging to this subculture influence your life?

2. Have you become more engaged in political issues through online sites? Why or why not? Why do you think so many young Americans take part in political activity both online and offline these days?

3. Can you recall a time when someone gave you discouraging feedback when you attempted to engage in a conversation about political issues? If so, why do you think the person discouraged you? How did it make you feel? How do you think you could have approached that conversation in a way that would have been better able to engage that person in a discussion about political issues?

4. Do you think political knowledge is valued on your campus? Provide examples to back up your answer. How often do you talk to others about political issues on campus? Why?

5. Think about the dominant culture in the United States. What are some of the dominant norms, values, and artifacts that (a) encourage or (b) discourage political participation by Americans? That encourage or discourage equitable views on race, gender, sexuality, nationality, various religions, ethnicity, and so forth?

6. Do you think recycling is a norm on your campus? What makes you think so? If it is a norm, what do you think brought about the addition of recycling to the campus culture norms? If recycling is not now a norm on your campus, why do you think that is? What could you do to help promote recycling on campus?

7. How much food do you think is wasted on your campus, including in the cafeterias? What value(s) does it represent in our culture that food waste is accepted? What can you do to work with your campus cafeteria and the students and others who spend time on campus to change this culture? How will you begin?

8. Why do you think Americans, in recent years, have become increasingly politically polarized? In general, how do you think such polarization influences nations with democratically elected governments? What do you think are some of the present and potential repercussions of political polarization on the well-being of the United States?

9. Nearly all actively researching climate scientists (generally considered to be 97 percent) and the vast majority of all scientists (80–92 percent range

from various studies) agree that global climate change is being partially caused by human behavior. In 2018, 8 out of 10 Republicans and nearly all Democrats believe the science as well. What conflicting values do you think compete with this nearing consensus, somehow stopping Americans and the United States from taking more serious action to lessen the pace and effects of climate change? What are some steps you can take to change the values and culture on your campus, in your community, and in society regarding this issue?

SUGGESTIONS FOR SPECIFIC ACTIONS

1. Brainstorm with two or three of your classmates as to how you might promote civic engagement within the existing campus culture. If you do not think it is possible to do this within the existing culture of your campus, how might you change the campus culture to promote civic engagement on campus? Act on these ideas.

2. Organize a campus debate about a civic issue that affects the life chances (the chances one has to improve one's social class position and quality of life) of students on campus. Analyze the campus culture to figure out how to most effectively promote and advertise the event (and do so!).

3. Conduct three focus groups (after obtaining proper institutional review board approval) that explore (a) how your fellow students define the ideal campus culture and (b) the actual dominant norms and values of the campus culture. Recruit interested members of the focus groups to come up with an action plan to disseminate and act on your findings.

   Please go to this book's website at http://study.sagepub.com/white6e to find further civic engagement opportunities, resources, peer-reviewed articles, and updated web links related to this chapter.

NOTES

1. Two of the most well-known books that supply information about the Peltier case are Jim Messerschmidt's *The Trial of Leonard Peltier* (1983) and Peter Matthiessen's *In the Spirit of Crazy Horse* (1983/1992).
2. The Church of Jesus Christ of Latter-day Saints banned polygamy in 1890.
3. For example, Eric Robert Rudolph, a member of the white supremacist Christian identity movement, carried out the 1996 Centennial Olympic Park bombing and other bombings that targeted gay and lesbian Americans and abortion clinics.

REFERENCES

Abramowitz, A. I. (2010). *The disappearing center: Engaged citizens, polarization, and American democracy*. New Haven, CT: Yale University Press.

American Indian Movement. (1972). *The trail of broken treaties*. Retrieved from http://www.aimovement.org/ggc/trailofbrokentreaties.html

Anderson, M., Toor, S., Rainie, L., & Smith, A. (2018). *Activism in the social media age*. Pew Research Center. Retrieved from http://www.pewinternet.org/2018/07/11/activism-in-the-social-media-age

Bellafante, G. (2018, October 31). Is it safe to be Jewish in New York. *The New York Times*. Retrieved from https://www.nytimes.com/2018/10/31/nyregion/jewish-bias-safety-nyc.html

British Broadcasting Corporation. (2017, April 11). Malala Yousafzai made youngest UN Messenger of Peace. Retrieved from https://www.bbc.com/news/uk-39562122

Brooks, D. (2005, August 11). All cultures are not equal. *The New York Times*. Retrieved from http://query.nytimes.com/gst/fullpage.html?res=990CE2DA143EF932A2575BC0A9639C8B63

Brown, D. (2018, October 14). 19 million tweets later: A look at #MeToo a year after the hashtag went viral. Retrieved from https://www.usatoday.com/story/news/2018/10/13/metoo-impact-hashtag-made-online/1633570002/

Cacioppo, J. T., Cacioppo, S., Gonzaga, G. C., Ogburn, E. L., & VanderWeele, T. J. (2013). Marital satisfaction and break-ups differ across on-line and off-line meeting venues. *Proceedings of the National Academy of Sciences, 110*(25), 10135–10140. Retrieved from https://www.pnas.org/content/110/25/10135

Center for Information and Research on Civic Learning and Engagement. (2016, November 9). An estimated 24 million young people voted in 2016 election. Retrieved from https://civicyouth.org/an-estimated-24-million-young-people-vote-in-2016-election/

Fair Trade Campaigns. (2018). *Universities*. Retrieved from http://fairtradecampaigns.org/campaign-type/universities

Federal Bureau of Investigation. (2018, October 15). *Hate crime statistics, 2017*. Retrieved from https://ucr.fbi.gov/hate-crime/2017

Greenhouse, G. (2014, April 24). In Florida tomato fields, a penny buys progress. *The New York Times*. Retrieved from http://www.nytimes.com/2014/04/25/business/in-florida-tomato-fields-a-penny-buys-progress.html?_r=0

Hamilton, C. (2014, March 7). Malala Yousafzai: "I could either not speak and die, or speak and then die. I chose the second one." *The Independent*. Retrieved from http://www.independent.co.uk/news/uk/home-news/malala-yousafzai-shines-among-the-stars-as-she-inspires-thousands-of-schoolchildren-9177643.html

Howard, J. (2018, March 9). The history of the "ideal" woman and where that has left us. Retrieved from https://www.cnn.com/2018/03/07/health/body-image-history-of-beauty-explainer-intl/index.html

Human Rights Campaign. (2018). Violence against the transgender community in 2018. Retrieved from https://www.hrc.org/resources/violence-against-the-transgender-community-in-2018

Kohut, A., Doherty, C., Dimmock, M., & Keeter, S. (2011, May 4). *Beyond red vs. blue: Political typology* (Pew Research Center for the People & the Press). Retrieved from http://www.people-press.org/files/legacy-pdf/Beyond-Red-vs-Blue-The-Political-Typology.pdf

Kohut, A., Doherty, C., Dimmock, M., & Keeter, S. (2012). *Trends in American values: 1987–2012: Partisan polarization surges in Bush, Obama years.* Pew Research Center for the People & the Press. Retrieved from http://www.people-press.org/files/legacy-pdf/06-04-12%20Values%20Release.pdf

Lubitow, A. (2015). Breast cancer activism: Learning to write collaboratively for social change. In *Sociologists in action on inequalities: Race, class, gender and sexuality.* Thousand Oaks, CA: Sage.

Lubitow, A., & Davis, M. (2011). Pastel injustice: The corporate use of pinkwashing for profit. *Environmental Justice, 4*(2), 139–144.

Matthiessen, P. (1992). *In the spirit of Crazy Horse.* New York, NY: Viking Press. (Original work published 1983)

Messerschmidt, J. (1983). *The trial of Leonard Peltier.* Boston, MA: South End Press.

Pai, S., & Schryver, K. (2015). *Children, teens, media, and body image* [Issue brief]. Retrieved from https://www.documentcloud.org/documents/2170615-children-teens-media-and-body-image.html

Paul, A. (2014). Is online better than offline for meeting partners? Depends: Are you looking to marry or to date? *Cyberpsychology, Behavior, and Social Networking, 17*(10), 664–667.

Potok, M. (2017, spring). The year in hate and extremism. *Intelligence Report, 162,* 36–62.

The Princeton Review. (2014). *The Princeton Review* gives 832 colleges green ratings. Retrieved from http://www.princetonreview.com/green/press-release.aspx

Southern Poverty Law Center. (SPLC). (2018). Hate map. Retrieved from https://www.splcenter.org/hate-map

Student/Farmworker Alliance. (n.d.). *Victory over Taco Bell.* Retrieved from http://www.sfalliance.org/tacobell.html

Learning How to Act in Society

Socialization

W ould you like to make someone smile? Smile at the person! It's almost impossible not to smile back when someone is smiling at you. A smile is one of the first symbols we learn to use to communicate with others.

Babies learn to smile by interacting with their caregivers, who (often) smile at them as they play with, change, and feed them (see Lockhart, 2011; Messinger, 2005; Venezia, Messinger, Thorp, & Mundy, 2004). They quickly learn that smiling can make their parents happy (and, in particular, happy with them). The authors learned this when babysitting their many nieces and nephews. They could even smile to get their exhausted aunt and uncle to smile and laugh while changing them at 3:00 a.m. After fumbling their way to the changing table, they would look down to see them beaming up with a big grin. Even those *very* sleep-deprived people couldn't resist the urge to smile back at the niece or nephew who had just awakened them from their slumber. How could they not? They always smiled and looked so happy to see them!

On a more somber note about learned behavior, have you heard the tragic story of Isabelle, a girl who was kept locked in a closet until she was six years old (Davis, 1947)? When she was finally released, her behavior was akin to that of an untrained animal. She didn't know how to speak, eat with utensils, use a toilet, or even smile. Her lack of interaction with other humans had prevented her from going through a process of *socialization*, the way in which we learn how to interact effectively in society. Isabelle's experience (and that of other children deprived of social interaction) indicates that human contact and social interaction are crucial for proper human development.

The Looking-Glass Self

Everyone around us influences our self-perception and our behavior in some way. The early symbolic interactionist Charles Horton Cooley (1902) described how others affect our self-image with the term *looking-glass self*. By this, he meant that we perceive ourselves based on how we *think* others see us. For instance, as professors, we often find ourselves experiencing the looking-glass self when teaching a class. It is part of the job of a professor to continuously discern whether or not students understand a lecture and are engaged with the subject matter. Often, when we look at our students and see blank expressions, we assume that this means the class does not understand what we are saying. We may react to this by re-explaining the concept or trying to find a new way to present it. But what if the students are exhibiting the blank expressions because the concept is simple and the professor has already spent too much time on it? (Or what if they look that way because it's how they look when they are watching television, and they have blurred the difference between a recorded broadcast and a live interaction?) What Cooley's looking-glass self demonstrates is that our actions and behaviors are reactions to our *interpretations* of the people, objects, and situations we encounter.

Cooley would say that what matters in the above scenario is *not* what the students are actually thinking but what the professors *think* the students are thinking. This interpretation leads to subsequent behaviors, in this case, reiterating a concept that the students have already understood! Or we may misinterpret someone's genuine compliment "Nice outfit!" as a sarcastic comment and stop wearing the outfit because we think that the person disliked it. Cooley's theory can best be summed up this way: "I am not what I think I am, and I am not what you think I am. I am what I think you think I am."

The Generalized Other

George Herbert Mead established a theory of social behavior to show how individuals' personalities are developed through social experience. According to Mead (1913), we develop a *self* (the part of our personality that is a combination of self-image and self-awareness) by (a) interacting with others through the use of symbols and (b) being able to see ourselves through the perspective of others. "It is out of this conduct and this consciousness that human society grows" (Mead, 1918, p. 578).

Eventually, through social interaction, we complete our creation of the social self by developing a sense of how we might be seen through the eyes of any person (the *generalized other*) who espouses the prevailing norms and values of the society in which we live. We find ourselves watching and judging our own actions through the eyes of this generalized other, even when no other person is present. If you are reading this book alone, start singing the words you read. Now think about how doing so makes you feel. Do you feel embarrassed? If so, you have just experienced the influence of the generalized other. You have interacted with others enough to know how most people would react to hearing you sing the words in this text. Most likely, you've adjusted your behavior accordingly and stopped singing.

We also see the generalized other when examining important societal issues, such as *internalized racism*. One of the authors once had a brilliant African American colleague who would speak about how internalized racism served as a constant barrier to him as he attempted to navigate through his career in academia. Even though he was often the most brilliant mind in the room and even more often the deepest expert in a particular area of study, he described often still feeling inferior when white colleagues raised questions or critiques of his work, even if intellectually he knew how to counter their critiques. Maybe you're thinking to yourself, "he had low self-esteem or lacked confidence"? But this phenomenon didn't happen to him when he was with his black colleagues, only with his white colleagues. So what was going on? Internalized racism isn't about self-esteem or confidence, it's woven into a system and history of racism and oppression. Effectively, it comes from the constant racism woven into our society and the power and privilege that white people knowingly and unknowingly regularly have and exert; it is the "conscious and unconscious acceptance of a racial hierarchy in which whites are consistently ranked above people of color" (Johnson, 2008). My colleague would describe that his generalized other often interpreted situations through this lens of internalized racism, making him feel inferior and less capable, and thus he found himself presenting this way to white colleagues.

Ultimately, as Mead showed, we internalize the awareness of how our behaviors *seem* socially. Our personal identities, therefore, contain our sense of who we are in society and how others perceive us. No self can develop without social interaction.

The Id, the Superego, and the Ego

Sigmund Freud developed a theory of personality in the early 1900s that has influenced almost all studies of human personality since then. A crucial part of his work was to explore the role of the *unconscious* in human identity and behavior. He broke down the unconscious into three sections: the id, the superego, and the ego. According to Freud (Freud & Strachey, 1949/1989), the *id* consists of people's innate desires and urges (which are primarily centered on instant gratification, especially through sex and violence). The *superego* is the part of our unconscious that has internalized the dominant norms and mores of society (particularly what constitutes "right" and "wrong" behavior). It represents our awareness of others, especially their reactions to ourselves. The *ego* works to balance the desires of the id with the moral impulses of the superego. The ego is the part of the self with which we choose to act or not act on our desires. For Freud, the ego is constantly testing our sense of what we want against our sense of what is expected of us. Freud (2005) believed that "it is impossible to overlook the extent to which civilization is built up upon a renunciation of instinct" (p. 85). We learn about the dominant norms of society and how to balance them with our innate desires through socialization with other human beings.

Primary Socializing Agents

Of course, some people make more of an impact on us humans than others. These people tend to constitute or come from one or more of the primary socialization agents in our society. Today in the United States, there are five primary socializing agents: family, peers, education, the media, and religion.

Family

We cannot choose our parents, but they do have a tremendous influence over who we become. For most of us, from the moment we are born, our family members—particularly our parents—are those with whom we interact first and for the greatest amount of time. Whether a parent constantly tells us that we are mean and stupid or repeatedly praises us for being so caring and brilliant (or anything in between) can largely determine how we view ourselves (as dumb or smart) and how we behave (meanly or kindly). Similarly, being encouraged to be persistent and to keep working at difficult things until we succeed or being scoffed at for failing to do things quickly affects how we will deal with challenging situations in the future. These lessons of socialization can influence what we become. For example, we are much more likely to put time and energy into our work if we believe that we are smart and capable and that persistence pays off than if we think we are neither bright nor professionally competent if we don't get it "right" the first time.

Socialization also teaches us what we want (values) and helps to give us a road map to achieve these things (norms). We are not *born* as authors, Muslims, leaders, Republicans, or _____ (fill in the blank). We *become* these things because our families and other socializing agents have socialized us into becoming them. Parents from different families in our country, town, or neighborhood, with different values and norms, socialize their children in different ways and into different sets of norms and values. Likewise, parents from different societies (with different values and norms) socialize their children in different ways. For example, parents in the United States are likely to teach their children to desire different things from what parents in other societies, such as Afghanistan, El Salvador, or even Canada, would teach.

Families also influence how we behave as public citizens. For example, young adults are more likely to vote if they have parents who vote and discuss political issues with them, and this is the case across social class lines (Cicognani, Zani, Fournier, Gavray, & Bom, 2012; McIntosh, Hart, & Youniss, 2007; Gidengil, Wass, & Valaste, 2016). Engaged parents often shape engaged young Americans. Youth activists and WE Charity, Me to WE, and WE Day founders Craig and Marc Kielburger discuss the ways in which their parents encouraged them to be global citizens from a young age, and they offer great insight into how to raise civically engaged young people in their book *The World Needs Your Kid: How to Raise Children Who Care and Contribute* (Kielburger, Kielburger, & Page, 2010).

Peers

Our peers also have a great effect on how we view ourselves, how we interact with others, and what we become. Who hasn't wanted to impress one's friends, and who hasn't at some point or the other taken on some of the values and norms of one's peer group, even if sometimes they have contradicted the values of one's family? As soon as we are able to interact with people outside our family, we begin to be influenced by peers (those similar to us in terms of age, social class, etc.).

The children you played with in your neighborhood, school, sports teams, and so forth, constituted your peer group as you were growing up. To varying degrees, our peers influence our self-perception and behavior in the same way our family does. It is through our associations with others who are "like us," our *peer reference group*, that we learn what people like us are like.

One website for a summer camp included a quote from a camper saying, "At school I'm a nerd, but here I'm cool." You have probably heard other stories of people growing up with two different peer groups (e.g., one that consists of classmates during the school year and one of camp mates or peers on an inter-town sports team). While researching this book, we became aware of one young man who remembers being perceived (and thinking of himself) as a cool kid and a jock when around summer friends and as a lonely loser when among school peers. As a result, he found himself acting out the roles attributed to him by the different groups. He kept to himself and was rather shy and aloof with his classmates, but he found himself leading his summer friends on wild adventures he never would have dreamed of pursuing during the school year. His self-perception actually became a *self-fulfilling prophecy*. He changed his behavior from one environment to the other to become the person he thought his peers saw. This young man's story is a great illustration of the looking-glass self. Both his self-perception and his behavior were strongly influenced by how he imagined his peers viewed him.

Likewise, our attitudes can also be influenced by those around us. For example, whether or not your roommate or partner is cheery or gloomy can influence your own outlook on life. A study at the university of Notre Dame indicated that some students who started college with a positive attitude but were assigned a roommate with a negative attitude were showing signs that they were at risk for depression six months later. On the other hand, others who began college with a gloomy perspective on life but were assigned cheerful and positive roommates had developed a more positive thinking style after six months (Haeffel & Hames, 2013).

The language we use or that others around us use to describe individuals or groups of people also has a powerful socializing effect on how we think of and act toward them. For instance, the dehumanizing language often used to describe immigrants has a proven negative effect on how immigrants are viewed by the larger population (Utych, 2017). So, what happens when a leader of a country uses dehumanizing language to describe a group of people? As President Trump has taken to using dehumanizing language since 2016, suggesting that Mexican immigrants "infest" our nation, that they are "animals," "rapists," and "murderers or thieves," what effect does this have in shaping people's opinions about immigrants, Mexicans, and even people from other Latin American countries?

The key point is that attitudes are developed and altered through interaction (Vedantam, 2013), including language and images, and that these interactions socialize us into our values and subsequent behaviors.

Education

Institutions have multiple roles. Our educational experiences play a major, though often somewhat hidden, role in our socialization process. Schools serve the primary purposes of giving us the tools we will need to operate in society and socializing us to adapt to and follow the values that society has deemed most important. Another goal of schools, which sociologists can uncover using their sociological eye, is to teach Americans a "hidden curriculum": to respect the rules of how our society operates, to not challenge the status quo, and to act in nondisruptive ways (ways that will not disrupt the dominant norms of society). For example, among the first things you learn in school are how to raise your hand, wait your turn, sit quietly, and follow directions. Many of these habits are useful skills that kids need to learn to interact with other people. Yet these same practices also reveal that schools teach the values of conformity and obedience to authority through the hidden curriculum (Gracey, 1967). School also teaches the basic values and beliefs of our dominant culture (as outlined in Chapter 4). What we learn about different racial, ethnic, gender, and social class groups through our schools and textbooks (or what we *don't* learn because the information in school curricula and texts is limited and carefully selected) influences how we perceive and judge these groups throughout our lives. It also helps us understand our own social status and role in society.

Dominant (or popular) values and beliefs determine, to a great extent, what is taught in our schools at any given time. It is important to remember that history is taught through the perspective of the dominant groups in our society. An examination of how the portrayal of U.S. history has changed over the years (as different racial or ethnic groups have gained power in the United States) reveals this fact. Comparing U.S. textbooks with those recounting the same events in European textbooks or Russian textbooks will also reveal this fact.

Media

The *media* refers to any instrument or "medium" that disseminates information on a wide scale. Among the various media outlets, television is arguably the most important socializing agent today. The average American spends more time than ever watching television (Nielsen, 2012; Ravichandran & de Bravo, 2010). In fact, the average U.S. child spends more than 35 hours each week watching television, a 2.2-hour increase since 2009 (Rothman, 2013). Indeed, the average child actually spends more time each year watching television than spent in school! During those hours, children are exposed to shows created by corporations to draw viewers to the advertisements they use to market their products before, during, and after the programs. Children's perspective of the world, therefore, is influenced to a considerable extent by a carefully selected set of images (in both the television shows and the advertisements) that are used to promote business for advertisers. In fact, marketers spend more than $12 billion annually on the over 40,000 commercials an average child will see each year (Wilcox et al., 2004). That's 320,000 commercials an average American will view just during our teen years! Whether or not the information, language, and images on television are accurate or fair, it is one of the major windows many Americans have on the world around them.

With children (and all Americans) watching so much television, what are the messages being sent that help socialize us into our values regarding race, gender, sexuality, age, religion, ethnicity, nationalism, military, economy, globalism, and so forth? In a nation increasingly segregated by race, ethnicity, and social class, television shows are often the only way some Americans are exposed to other groups of Americans. Our perceptions and likely values (remember the looking-glass self and the generalized other) are influenced by the way these groups are portrayed in the media.

Religion

Americans are, overall, much more religious than citizens of other wealthy nations. Whereas 44 percent of Americans attend church once a week (this does not include funerals, christenings, and baptisms), 27 percent of people in Great Britain, 21 percent in France, 4 percent in Sweden, and 3 percent in Japan attend. Americans are—and by far—the most religious of all of the wealthy nations and all of the highly democratic nations in the world.

However, the number of Americans unaffiliated with a religious group has climbed rather dramatically over the past few years. Today, approximately 20 percent of Americans and 33 percent of 18- to 24-year-olds are religiously unaffiliated. Approximately two thirds of the unaffiliated say that they believe in God. "Overwhelmingly, they think that religious organizations are too concerned with money and power, too focused on rules and too involved in politics" (Pew Research Center's Religion & Public Life Project, 2012). These Americans are increasingly skeptical about the role of organized religion in society and in part are questioning the values that organized religion teaches around issues such as sexuality, gender, charitable giving versus social justice work, immigration, politics, and so forth. Through this example, we can see how young people have an awareness (right or wrong) about the *socializing agent* of organized religion and are questioning many of the values it teaches, while also holding closely to many of the values of being religious. Since historically one of the roles played by religion has been to encourage civic engagement in young people, it has become more important than ever that schools and universities provide civic engagement opportunities for students. Sociology courses offer one of the best vehicles for participation in society, as seen in the Sociologists in Action box.

Sociologists in Action

Donna Yang, Christian Agurto, Michelle Benavides, Brianne Glogowski, Deziree Martinez, and Michele Van Hook

In the following paragraphs, students at William Paterson University describe how they used the sociological tools they gained in their Principles of Sociology course to educate their classmates about Kiva, a microfinance project designed to aid women and diminish poverty globally, and to participate in the organization's efforts.

After enrolling in an introductory sociology course at William Paterson University, we collaborated to complete a group assignment that was designed to utilize various sociological concepts developed within the classroom. We designed a project after learning about Kiva, an organization dedicated to helping ordinary citizens like us to become lenders to entrepreneurs throughout the poorer areas of the world who are seeking microfinance. We sought not only to raise money toward a Kiva loan, but—more important—to generate awareness about micro-finance's ability to empower women and help fight poverty. The website, www.kiva.org, is designed to facilitate partnerships between everyday people with a passion to give and low-income entrepreneurs seeking financial services. Kiva partners with a microfinance institution to provide small loans, typically ranging from $25 to $3,000, that are temporarily financed by Kiva users who have browsed the profiles of potential entrepreneurs uploaded on the website.

Although gender inequality was central to our class project, there were many other key sociological issues that were incorporated within its framework. The purpose of the assignment was to promote the application of the two "core commitments" of sociology—using the sociological eye and promoting social activism. Utilizing our collective sociological eye, we were able to look beneath the surface of society and recognize patterns of inequality intersecting along both class and gender lines. We realized that social stratification has created hierarchies through which women are marginalized economically, politically, and socially. Microfinance programs such as Kiva seek to mitigate this unequal access to wealth by creating greater accessibility for low-income individuals to financial services. In this way, they help to redistribute the accessibility of wealth across a wider spectrum of social classes.

As a group, we felt passionate about Kiva because of its commitment to improve society by providing more equal lending opportunities. We raised more than $150 toward a group loan for the communal bank Mujeres Progresistas (Progressive Women) in Cuenca, Ecuador. This bank is composed of 11 women working to improve their household economic situations through their jobs. Our specific loan was distributed to 2 women—both mothers of four—seeking loans to help finance their personal businesses in order to help support their families. Maria Huerta, who lives in the San Joaquin area, will use this loan to invest in buying chickens and chicken feed to begin running a chicken farm. Maria Suqui, who lives in the Feria Libre area, will be using the loan to help establish a snack shop, through which she hopes to earn enough money to eventually own a home.

(Continued)

(Continued)

In order to promote awareness about our project and Kiva's initiatives, we also created informational pamphlets and distributed them to students on our college campus. The pamphlets explained microfinance, how Kiva utilizes this process of lending to help low-income individuals around the globe, and how students can get involved. Additionally, a PowerPoint was presented to peers enrolled within the Principles of Sociology class in hopes of inspiring other peer members to join Kiva's efforts. This project helped us to realize the importance of sociology, its applicability within everyday life, and our obligation to act as socially conscious individuals in order to help promote greater social justice, including gender justice.

Source: Courtesy of Donna Yang.

Exercise 5.1
Family Influence and Civic Engagement

Think about how your family has influenced your level of civic engagement.

1. Did your parents and other family members discuss political issues with each other when they were around you (at the dinner table, over coffee, etc.)?

2. Did your parents and other family members discuss political issues with you when you were growing up? Do you discuss political issues with them now? Why or why not?

3. If you answered yes to Question 2, have they asked you to participate in political activities with them? Have you? If so, in what ways? How did those experiences influence your own view of the importance of civic engagement?

4. If your parents and other family members are not active citizens, why do you think they are not? What do you think might encourage them to become active?

5. Overall, how do you think your parents' and other family members' level of civic engagement affected your own level of civic engagement?

6. After taking a moment to really evaluate the implications of this, describe how you feel about your own level of civic engagement.

7. How will you influence your own children's political participation (when or if you have kids)?

Exercise 5.2
Your Many Selves

Everyone acts (at least somewhat) differently among different groups of people. Write a one-page essay describing different situations in your life in which you act almost as though you are different people (different versions of yourself). Why do you act differently in these different situations? What role do you think the generalized other (Mead) and the looking-glass self (Cooley) play in socializing you into these different roles? Be sure to define these terms in your paper before you give examples of them, and dig beneath the surface in your analysis.

Exercise 5.3
Peers and Civic Engagement

Think about how your peers have influenced your level of civic engagement.

1. Do your peers discuss political issues with each other when they are around you?

2. Did your peers discuss political issues with you when you were growing up? Do you discuss political issues with your peers now? Why or why not?

(Continued)

(Continued)

3. Do your peers vote? Have you ever gone to the polls with a peer? If so, what was that like?

4. Are your peers civically active? If yes, in what ways? Have they asked you to participate with them? Have you? If so, in what ways? How did those experiences influence your own view of the importance of civic engagement? If they have asked you to participate with them and you didn't, why didn't you? What do you think you might have lost out on by not participating? If your peers are not active citizens, why do you think they are not? What do you think might encourage them to become active?

5. Overall, how do you think your peers' level of civic engagement has affected your own activism?

Extra credit: Read "Young Voters Could Make a Difference. Will They?" (https://www.nytimes.com/2018/11/02/us/politics/young-voters-midterms. html). Having read this, does it change your perceptions about (a) the importance of voting, (b) whether you might vote in the upcoming elections, or (c) the ways you and your peers talk (or don't talk) about politics?

Exercise 5.4
The Hidden Curriculum: History, Values, and Socialization

Find a U.S. history book currently used at a local public grammar, middle, or high school.

1. Note the title of the book. What image does the title convey? Is there a photograph or illustration on the cover of the book? If so, what message does it convey?

2. Look at the table of contents. What topics are covered? Think of five additional chapters you think could or should have been included in the book. Why do you think they could or should

have been included? Why do you think they were not included? What does your study of this textbook tell you, sociologically, about the values that the school's curriculum indicates are most important? Besides the need to keep textbooks a certain length, why else do you think certain topics have not been included?

3. Compare how different racial, ethnic, gender, religious, and social class groups are covered (or not covered) in the book.

4. Compare the Civil War chapter in the history book you obtained with Howard Zinn's chapter on the Civil War in his book *A People's History of the United States, 1492–Present* (2003). Your school library should have a copy, or you can find it online (https:// mvlindsey.files.wordpress.com/2015/08/peoples-history-zinn-1980. pdf). (Read Chapter 10.) What are the similarities? What are the differences? How do you think the differences relate to the "hidden curriculum" of the public school system in the United States?

5. How do you think members of different races and ethnicities, genders, and social classes would view their respective status positions and roles in society after reading each of the two chapters? What might they say is missing? What might they want to add?

Exercise 5.5
Coverage of Social Activism on Television News Programs

For this assignment, you will be conducting a content analysis of 10 news programs. Your focus will be on how these news programs cover social activism. Make 10 copies of the Content Coding Sheet (Figure 5.1). You will fill these charts in as you view your news programs. As you fill out these charts, there are a few rules:

1. You must view news programs during the same time slots (e.g., early morning, midmorning, afternoon, prime time, late night) on

(Continued)

(Continued)

one particular channel (e.g., ABC, NBC, CNBC, CNN, Fox News, CBS, PBS, BBC).

2. All 10 programs should be either local or national and international.

3. After having filled out your Content Coding Sheets, you will fill out your Data Summary Sheet (Figure 5.1). This will give you the overall numbers and percentages.

4. Once you have completed the data collection, join with three other members of your class to analyze your collective data (making sure that not all the members of your groups analyze the same news programs). What do your results tell you about how social activism is portrayed on the news programs you examined? What does this tell you about how we are socialized to view social activists and social activism? Why do you think the media portrays activism this way? What is missing from the portrayals of activists and activism? Is there a "hidden curriculum" in what we learn (and don't learn) from the media?

Figure 5.1   Content Coding and Data Summary Sheet

Content Coding Sheet

Type of social activism: _____

Name of news program: _____

Station viewed on (provide channel and network): _____

Local or national/international: _____

Time of day (a.m. or p.m.): _____

Note the number of examples of social activism covered and note the type—for example, (a) violent or peaceful, (b) disruptive or nondisruptive—and how it is portrayed (positively or negatively) by checking under the appropriate columns.

Examples of Activism	Violent	Peaceful	Disruptive	Nondisruptive	Positively	Negatively
1.						
2.						
3.						
4.						
5.						

Data Summary Sheet

News program: _____

Station/network: _____

Local or national/international: _____

Time of day: _____

Type of Activism	Number Portrayed	
	Positively	Negatively
Violent		
Peaceful		
Disruptive		
Nondisruptive		
Total		

Note: Reproduction authorized only for students who purchase this book.

Exercise 5.6
Young Voters Matter

Listen to "Young Voters," an interview with Andy MacCracken, the executive director of the National Campus Leadership Council, at https://www.c-span.org/video/?311416-4/young-voters, and then answer the following questions:

1. What were the main points discussed in this video? Were you aware of them before? Why or why not?

2. What questions from callers particularly interested you? Annoyed you? Why? How do they relate to the ideological divide in politics today?

3. Based on both your own experiences and what you learned in this video, if you were to advise a presidential candidate on how to turn out the college student vote, what would you suggest that he or she do?

4. Next, read "High School Civic Education Linked to Voting Participation and Political Knowledge, No Effect on Partisanship or Candidate Selection" (https://civicyouth.org/high-school-civic-education-linked-to-voting-participation-and-political-knowledge-no-effect-on-partisanship-or-candidate-selection/).

5. Based on this article, what is your analysis of the importance of civic education both to the civic act of voting and to all forms of civic engagement?

DISCUSSION QUESTIONS ─────────────

1. Think of the five primary socializing agents. Now rank order them from the one that influenced you the most (#1) to the one that influenced you the least (#5). Now, explain why you ranked them this way. How deeply, however, do you think that #4 and #5 also influenced you, including your values and norms?

2. Think about Cooley's concept of the looking-glass self. Do you see yourself as someone who can make a positive impact on society? Why

or why not? In what ways has your own looking-glass self experience influenced this? If you answered "no" that you don't see yourself as someone who can make a positive impact on society, how does analyzing your socialization experience help you to start to think differently, that there is a possibility for resocialization of your thinking on this?

3. Why do you think more young people are moving away from organized religion? If this trend continues, how and where do you think society can fill in the positive values young people still associate with religion while eliminating the negative values young people are increasingly associating with religion? Or do you agree with Marx that religion is an "opiate" distracting us from organizing for real, structural social change?

4. Which primary socializing agents could most effectively foster civic engagement across the United States? Why? How? What are the ways they are currently doing so, and in what ways can they improve? Be specific and well thought-out in your suggestions.

5. What worldwide socializing agents might effectively encourage civic engagement across the globe?

6. Which of the following sociological theories—conflict, functionalist, or symbolic interactionist—would you find most useful if you were to conduct a study on political participation in the United States? Why?

7. Have your values and norms about environmental issues, such as recycling or global warming (or the views of someone close to you), changed in recent years? If so, what were the social forces that led to this resocialization process? What does this tell you about the possibility of resocialization (for you and for society) on other seemingly embedded ways of thinking (values and norms)?

8. Being honest, what are the values and norms you were socialized into regarding race? Sexuality? Immigrants? Homeless individuals? Who and what socialized you to these values, and why did they choose this set of values? Which, if any, of these values would you hope to see yourself reexamining, being open to resocialization toward a more inclusive or equitable set of values?

9. When you became resocialized after starting college, did your political activism increase, decrease, or remain the same? Why? How does the culture of your college affect your response to this question?

10. How do you think we can create a society in which citizens are better socialized to become knowledgeable and effective participants and agents of social change?

SUGGESTIONS FOR SPECIFIC ACTIONS

1. Determine whether or not your college or university has committed itself to the educating citizens movement and whether it has taken steps to teach its students to become knowledgeable, effective citizens. One way to do this is to find out if your president or provost has signed the Campus Compact pledge (found at https://compact.org/actionstatement/), joined the American Democracy Project (http://www.aascu.org/programs/adp), or signed on to some other such initiative to promote civic engagement in higher education. If he or she has, then compare the goals of the initiative with the results you see on campus. Write a letter to your provost or president and your school newspaper that includes the results of your research. If your school has not embarked on such a mission, write a letter to your provost or president explaining what the educating citizens movement is about (after having looked at the Campus Compact and American Democracy Project websites) and asking what steps your school has taken to encourage well-informed and effective citizens.

2. Explore the institutional connections your college or university has established with the surrounding community (e.g., service-learning courses, school-sponsored community centers, school-sponsored tutorial programs, research partnerships between professors and community members, etc.), and write a letter to your local paper that summarizes your key findings (to help inform the campus and community about these matters).

3. Participate in one of the events your university uses to connect to the local community (sponsoring a community fair, volunteering to work at a local school, etc.). Use your sociological eye to examine the power dynamics behind how the event or program is coordinated through the school and community. For example, who designed the program (the school, community, or both)? Who coordinates the program? Who has authority over the program? Who funds the program? How do you think this type of coordination influences the overall program? Explore the idea of "partnership" in your analysis and how to ensure true partnerships (with the community, the organizations doing the work, and the individuals being helped) are driving the work.

4. Set up an interview with the person who coordinates the orientation program for first-year students at your university. Inquire as to (a) how orientation socializes students into successfully learning and adopting the campus rules, norms, and values and (b) what mechanisms are used during the orientation process to socialize new students into the role of educated and engaged citizens. If no such mechanisms exist (or the ones that do exist are inadequate), offer some suggestions about what could

be put into place to the person who coordinates the first-year orientation program.

Please go to this book's website at http://study.sagepub.com/white6e to find further civic engagement opportunities, resources, peer-reviewed articles, and updated web links related to this chapter.

REFERENCES

Center for Information and Research on Civic Learning and Engagement. (2013, January 17). High school civic education linked to voting participation and political knowledge, no effect on partisanship or candidate selection. Retrieved from https://civicyouth.org/high-school-civic-education-linked-to-voting-participation-and-political-knowledge-no-effect-on-partisanship-or-candidate-selection/

Cicognani, E., Zani, B., Fournier, B., Gavray, C., & Bom, M. (2012). Gender differences in youths' political engagement and participation: The role of parents and of adolescents' social and civic participation. *Journal of Adolescence*, *35*(3), 561–576.

Cooley, C. H. (1902). *Human nature and the social order*. New York, NY: Schocken Books.

C-SPAN. (2013). Young voters. Retrieved from https://www.c-span.org/video/?311416-4/young-voters

Davis, K. (1947). Final note on a case of extreme isolation. *American Journal of Sociology*, *52*(5), 432–437.

Ember, S. (2018, November 2). Young voters could make a difference. Will they? *New York Times*. Retrieved from https://www.nytimes.com/2018/11/02/us/politics/young-voters-midterms.html

Freud, S. (2005). *Civilization and its discontents*. New York, NY: W. W. Norton.

Freud, S., & Strachey, J. (Eds.). (1989). *The ego and the id* (J. Riviere, Trans.). New York, NY: W. W. Norton. (Original work published 1949)

Gidengil, E., Wass, H., & Valaste, M. (2016). Political socialization and voting: The parent-child link in turnout. *Political Research Quarterly*, *69*(2), 373–383.

Gracey, H. L. (1967). Learning the student role: Kindergarten as academic boot camp. In D. Wrong & H. L. Gracey (Eds.), *Readings in introductory sociology* (3rd ed., pp. 215–226). New York, NY: Macmillan.

Haeffel, G. J., & Hames, J. L. (2013, April 16). Cognitive vulnerability to depression can be contagious [Online]. *Clinical Psychological Science*. doi: 10.1177/2167702613485075. Retrieved from http://cpx.sagepub.com/content/early/2013/04/15/2167702613485075.full.pdf+html

Johnson, R. (2008). *The psychology of racism: How internalized racism, academic self-concept, and campus racial climate impact the academic experiences and achievement of*

African American undergraduates (Doctoral dissertation). Retrieved from ProQuest Dissertations & Theses database.

Kielburger, C., Kielburger, M., & Page, S. (2010). *The world needs your kid: How to raise children who care and contribute*. Vancouver, BC, Canada: Greystone Books.

Lockhart, S. (2011). Active learning for infants and toddlers. *Resource, 30*(1), 5–10.

McIntosh, H., Hart, D., & Youniss, J. (2007). The influence of family political discussion on youth civic development: Which parent qualities matter? *Political Science & Politics, 40*(3), 495–499.

Mead, G. H. (1913). The social self. *Journal of Philosophy, Psychology, and Scientific Methods, 10*, 374–380.

Mead, G. H. (1918). The psychology of punitive justice. *American Journal of Sociology, 23*, 577–602.

Messinger, D. (2005). A measure of early joy? Afterword to the republication of "All Smiles Are Positive, but Some Smiles Are More Positive Than Others." In P. Ekman & E. Rosenberg (Eds.), *What the face reveals: Studies of spontaneous facial expression using the Facial Action Coding System (FACS)* (2nd ed., pp. 350–353). New York, NY: Oxford University Press.

Nielsen. (2012). *Television measurement*. Retrieved from http://www.nielsen.com/us/en/measurement/television-measurement.html

Pew Research Center's Religion & Public Life Project. (2012, October 9). *"Nones" on the Rise*. Retrieved from http://www.pewforum.org/2012/10/09/nones-on-the-rise/

Ravichandran, P., & de Bravo, B. F. (2010, June). *Young children and screen time (television, DVDs, computer)* (National Research Center for Women and Families). Retrieved from http://www.center4research.org/2010/05/young-children-and-screen-time-television-dvds-computer

Rothman, L. (2013, November 20). FYI, parents: Your kids watch a full-time job's worth of TV each week. *Time*. Retrieved from http://entertainment.time.com/2013/11/20/fyi-parents-your-kids-watch-a-full-time-jobs-worth-of-tv-each-week/

Utych, S. M. (2017). How dehumanization influences attitudes toward immigrants. *Political Research Quarterly, 71*(2), 440–452.

Vedantam, S. (2013, June 24). *Gloomy thinking can be contagious*. Retrieved from http://www.npr.org/blogs/health/2013/06/24/193483931/Contagious-Thinking-Can-Be-Depressing

Venezia, M., Messinger, D. S., Thorp, D., & Mundy, P. (2004). The development of anticipatory smiling. *Infancy, 6*(3), 397–406.

Zinn, H. (2003). *A people's history of the United States, 1492–present*. New York, NY: HarperCollins.

Wilcox, B. L., Kunkel, D., Cantor, J., Palmer, E., Linn, S., & Dowrick, P. (2004). *Report of the APA task force on advertising and children*. Washington, DC: American Psychological Association.

Zinn, H. (2003). *A people's history of the United States: 1492–present*. New York, NY: Harper.

CHAPTER 6

Deviant Behavior and Social Movements

W hat would be your reaction if you saw a man sitting in his office, dressed in knee-length pants, wearing a wig, and snorting tobacco? Would you think his behavior is deviant? Perhaps! However, sniffing tobacco and wearing a wig and breeches were once considered normal for men. In fact, in the colonial United States, engaging in such behavior demonstrated high social status. Our understanding of what is deviant behavior is socially constructed and, therefore, changes over time and from society to society.

Although behaviors such as incest and killing innocent people are considered deviant in almost all societies, much else of what is considered deviant behavior varies across cultures and time periods. Notions of what constitutes deviant behavior can also be situational. There are circumstances in which acts that would normally be considered completely beyond the bounds of acceptable behavior, including many brutal acts of violence, can come to be seen as normal and appropriate. For example, soldiers' reports and memoirs frequently include accounts of acts that may seem unjustifiable once the fighting has ended but had seemed acceptable, even necessary, in the midst of it. This chapter will discuss (a) how deviance is defined, (b) how the three major theoretical perspectives and other theories view deviance, (c) why some social groups are labeled as deviant, and (d) how some social movements have changed society by "normalizing" behavior that had previously been thought of as deviant.

As you will recall from Chapter 4, cultural norms are socially constructed expectations for behavior. A society's norms include basic behaviors, manners and etiquette, and laws and legal codes and follow the society's unique culture and ways of doing things. As members of a society, we learn which behaviors are appropriate through socialization. Our actions are guided by how we see others behave, what we are taught by our socializing agents, and the reactions of those with whom we interact.

Folkways are norms that are enforced through *informal rewards* and *sanctions* (e.g., approval or disapproval from others). For example, we follow folkways when we hold the door for someone behind us, let people we speak with finish their sentences without interruption, and make conversation with people we meet at parties. Violating these folkways might result in a dirty look or being considered rude, socially awkward, or strange. However, this type of deviant behavior hardly ever results in serious social consequences and never results in legal repercussions.

Mores, on the other hand, are those norms related to behaviors that reflect the values society holds most dear. Abiding by social and legal contracts, wearing clothing in public, respecting other people's property, and adhering to certain sexual behaviors are all examples of mores. Violating these norms results in much stronger reactions and, perhaps, criminal charges (many mores are supported by laws). For example, refusing to support a colleague in need, cheating on a girlfriend or boyfriend, or openly announcing you are bisexual often results in people expressing strong reactions that can have negative repercussions (e.g., being labeled as "not a team player" at work, being labeled as untrustworthy and selfish by friends, or being thought of as sexually deviant by some, although the latter is a good example of a rapidly changing more, whereby society is becoming more accepting of this as a norm).

Violations of our most powerful mores, those representing our deepest values (e.g., confining sexual relations to those not closely related, limiting our meat eating to animals rather than other humans), are called *taboos*. Incest—sexual relations between close family members—is a taboo in almost all societies. Cannibalism—the practice of eating humans—is also widely considered to be a taboo. Those who carry out such acts are regarded as socially repugnant and face widespread, if not universal, condemnation from the other members of their society.

Laws are norms that are formalized rules of behavior enacted by legislatures and enforced by formal sanctions carried out by the criminal and civil justice systems. Laws provide guidelines as to what people should do (e.g., drivers should stop at stop signs, citizens should pay taxes, parents should send their children to school) and what they should *not* do (e.g., commit rape, robbery, or insurance fraud, or possess or use certain drugs). Those who break laws are subject to fines, imprisonment, and possibly even capital punishment.

Functionalism and Deviance

Émile Durkheim (1895/1982) established the functionalist perspective on deviance. He maintained that deviant acts and those who commit them actually serve several useful and necessary purposes or functions for society. Deviants can help make the norms of society clearer to the majority population, unite the nondeviant members of society, and even promote social change.

Sometimes, it is difficult to know what the norms of society are until we see someone punished for violating one. When some members of society are punished for committing deviant acts, the rules of society are clarified and reinforced for everyone. For example, seeing someone pulled over for speeding usually makes other drivers slow down. Observing a classmate punished for plagiarism helps students recognize that stealing someone else's words or ideas is unacceptable behavior and that those who do so are liable to be punished. We already know that speeding and plagiarism are wrong, but seeing these mores violated reminds us of how important it is to abide by these rules and often (if only through fear of being punished) causes us to act as society wishes.

Bridging and Bonding Social Capital

According to Durkheim (1895/1982), deviants can also unify members of society. When people see deviant behavior as a threat, they come together to punish and root out the deviant members of society. United, they can create *social capital*, a network of interconnected people who can rely on one another and work together to accomplish a goal. For example, in the mid-1990s, the parishioners of St. Patrick's parish in Brockton, Massachusetts, became closely united when they organized to confront a prostitution and drug operation headquartered in abandoned buildings across the street from the church. The criminal activity centered in the abandoned buildings made the parishioners fearful when they came to church, and they were tired of being afraid. They knew that they really had to do something when the pastor was propositioned by a prostitute on the way to Mass! Working together, they compelled the city to take over the houses (in lieu of back taxes owed), raze them, and drive out those using them for deviant acts that made the parishioners fear going to church.

The experience of eliminating the individuals exhibiting deviant behavior from their church neighborhood *bonded* the parishioners more closely, as they learned that they could rely on one another and become a powerful force when united. The members of St. Patrick's parish were also able to make connections across the city of Brockton and join with the members of different religious organizations to address safety issues, such as crime, that all Brocktonians faced. In doing so, they showed the power of what is known as *bridging* social capital. They were able to make connections (bridges) among different groups in Brockton that enabled them to begin to know and work with one another effectively (social capital).

The Me Too Movement is a powerful example of citizens uniting to bring attention to sexual assault and sexual harassment, issues of deviance that are pervasive in U.S. society. While it is a clear cultural more that "no means no" and that sex is to be a mutually agreed upon act, 33 percent of women and 17 percent of men in the U.S. have experienced sexual violence during their lives, and 20 percent of women and 1.5 percent of men have experienced rape or attempted rape (NSVRC, 2019; Smith et al., 2017). Another deeply held more in our society is that no person under 18 should be forced to have sex with an adult, and yet 7 percent of girls (8.5 million women) and 1.3 percent of boys (1.5 million men) in the United States experienced rape when they were under the age of 18. The Me Too Movement began in 2006 as a way to help women who have survived sexual violence and gained great momentum in 2017 and 2018 when the hashtag #MeToo gained millions of hits and became a space for women (and some men) to speak out about their experiences as sexual assault survivors. By the end of 2018, one manifestation of the movement was that over 200 men in powerful positions, identified as perpetrators of sexual assault or sexual harassment, had lost their jobs (Carlsen et al., 2018). This is an example of deviant behavior allowed to become a norm in our society and now being challenged by a powerful and effective social movement.

Conflict Theory and Deviance

As you will recall from Chapter 2, Marxist and other conflict theorists maintain that society is made up of groups competing for power. In turn, they believe that norms and laws are largely created by and for the benefit of those who hold the most power in society. Marx argued that the members of the ruling class (the owners of the means of production) use every instrument at their disposal, including shaping norms and laws, to protect their interests. Therefore, the actions of the poor are more likely to be labeled as deviant and criminal than those of rich individuals or major corporations.

In *The Rich Get Richer and the Poor Get Prison*, Jeffrey Reiman and Paul Leighton use the conflict perspective to argue that the criminal justice system is biased against poor people and is in favor of wealthy individuals and corporations (Reiman & Leighton, 2017). They maintain that corporations actually do much more harm to society by committing acts that are not officially labeled "criminal" than do individuals by committing acts that are considered crimes. For example, unsafe workplaces (e.g., leading to black lung disease, asbestos-based cancer, repetitive-motion injuries, or preventable deaths or injuries due to cave-ins in coal mines), unsafe medical practices (e.g., unnecessary surgeries, lack of proper medical care, unsafe medical devices such as mesh, prescribing combinations of drugs leading to drug interactions), environmental pollution (e.g., from cigarettes, from chemical and industrial factories), and lax oversight of consumer safety (e.g., defective tires, tainted meat) harm far more people than criminal acts carried out by individuals. However, they are relatively ignored by the judicial system, whereas the crimes of predominantly poor and minority citizens result in disproportionately severe punishments.

Symbolic Interactionism and Deviance

Symbolic interactionists maintain that people learn to conform to or deviate from the norms of society through their interactions with others. They stress that deviance is socially constructed and that deviant behavior is learned. Two of the most well-known theories on deviance that fall under the umbrella of the symbolic interactionist perspective are differential association theory and labeling theory.

Through his *differential association theory*, Edwin Sutherland (1947) maintained that people learn deviant behavior by associating with people who commit deviant acts. People tend to base their own behavior on the behavior of those with whom they interact. Therefore, those who are raised in families that violate social norms or spend much of their time with friends who break the norms of society will be more likely than others to act in deviant ways. For instance, Jonathan was involved in a mentoring program, Friends and Mentors for Change (FAM for Change), for inner-city youth who are at high risk of not graduating from high school. His initial research results revealed that when these kids, who face negative situations at their schools or homes, are surrounded by external forces that may influence them in negative ways, they are much more likely to commit deviant acts than when they are spending time in the FAM for Change program, surrounded by more positive role models. Statistically, the students were all considered at risk of not graduating high school, with a predicted dropout rate of 75 percent based on a set of variables determined by their high school educators. With the work of the mentoring program, however, 100 percent of the first two cohorts in the program graduated high school! The DREAM mentoring program is also a great example of college students starting an organization to create social change (https://www.dream program.org/). Established by students in 1999 at Dartmouth College in New Hampshire, it now has brought its Village Mentoring model to Vermont, Massachusetts, and Pennsylvania, creating positive change in the lives of both mentees and their college mentors.

Labeling theory, another symbolic interactionist perspective on deviance, focuses on the categorization of people as deviant. Howard Becker (1963), the founder of labeling theory, maintained that behaviors and individuals *become* deviant only when people with some social power *label* them as deviant. Once people so labeled accept this categorization, they begin to consider themselves deviant and act accordingly. For example, when a young woman becomes known to others as a drug addict, is ostracized by those who do not use illegal drugs, and is given the label "druggie," she begins to socialize more and more with fellow druggies and act like the druggie that people perceive her to be. We can see the effects of this on high school campuses when kids who have been given a label that makes them outcasts (e.g., "loser," "nerd," "burnout") begin to take on the qualities of that label, sometimes with violent results.

Labeling Some Social Groups as Deviant

As noted above, deviance is a *social construction*, which means that behaviors are not in themselves normal or deviant; they become deviant only when society defines them as such. Unfortunately, it is easy and common to define or perceive entire groups of people as deviant rather than merely defining specific behaviors as deviant acts. For instance, *racial profiling* is a controversial practice in law enforcement that involves subjecting people to greater scrutiny or treating some people as potentially dangerous based on stereotypes and prejudices associated with their perceived race.

The state of New Jersey became a focal point for a national debate on racial profiling in the 1990s. Black motorists complained, and an independent review confirmed that they were being pulled over for routine traffic stops and even searches far more often than members of other racial and ethnic groups and often without cause (Lamberth, 1994). A whole new term has even entered our collective lexicon to describe this unfair treatment: *driving while black*. A series of reviews followed, including a number of hearings in the New Jersey State Senate, resulting in the establishment of guidelines for law enforcement officers that point out that "any officer can unwittingly or subconsciously fall prey to racial or ethnic stereotypes about who is more likely to be involved in criminal activity" (New Jersey Division of Criminal Justice, n.d., para. 5) and mandate that officers work to counter this fact by focusing on people's conduct rather than their racial or ethnic appearance (New Jersey Division of Criminal Justice, 2009). Police efforts still seem to disproportionately focus on race. For example, the New York City Police Department's "stop and frisk" policy seems to target racial minorities. In 2011, 87 percent of the 684,330 people stopped and questioned on the street by New York City police officers were either black or Hispanic. "Ten percent of the stops led to arrests or summonses and 1 percent to the recovery of a weapon" (Elignon, 2012, para. 15). In 2013, a judge declared the practice unconstitutional, saying that "the city adopted a policy of indirect racial profiling by targeting racially defined groups for stops based on local crime suspect data" (Goldstein, 2013).

Deviance and Organizing for Change

It is useful for the well-being and maintenance of society to construct norms and to recognize forms of deviance. Without these, people would need to navigate the basic rules of behavior in every situation they enter. How inefficient and chaotic that would be! However, social norms that reinforce the worldviews and interests of dominant groups (those in power) often act as repressive forces against the worldviews and interests of minority groups and all those who choose to deviate (even in harmless ways) from the dominant culture. Once a perspective takes on the veneer of "reality" or "truth" or "the right way," it is difficult to change. Usually, only concerted, organized effort can effectively counter such social forces. And the history of the United States is full of people doing just that—coming together to create the changes they desire for society.

Society needs people willing to break norms and mores (even those upheld by laws!) that are unfair, unjust, or inequitable. U.S. and world history is replete with examples of these types of *social deviants*, such as women fighting for equal rights; American revolutionaries (before the war for independence was won); Gandhi and his movement to overthrow the caste system in India; Nelson Mandela, Desmond Tutu, Stephen Biko, and the anti-apartheid movement in South Africa; and civil rights leaders, who became heroes by protesting against and helping overturn unjust laws and governments. For example, Mildred Jeter, a part black part Native American woman, and Richard Loving, a white man, were considered criminals when they married in 1958, in violation of the law in their state of Virginia that prohibited interracial marriage. However, their protest against this unjust law resulted in the 1967 Supreme Court decision (*Loving v. Virginia*) that declared such laws unconstitutional, leading to a tremendous increase in the number of interracial marriages, with 17 percent of all U.S. marriages in 2015 being interracial (Pew Research Center, 2018).

The U.S. civil rights movement is probably the most famous example of people performing socially deviant acts to organize for social change. Before that movement, segregation laws in many areas of the country pro- hibited black Americans from using the same facilities or drinking from the same water fountains as white Americans. Under the "separate but equal" laws, blacks were relegated to inferior schools. During the 1876 through 1965 "Jim Crow" era in the South, local and state laws in that part of the country mandated racial segregation. Black people were harassed when attempting to vote during the period as well. Some were the victims of violence, and others were simply turned away without being allowed to cast their votes. Jim Crow laws also imposed a large variety of other restric- tions on black people that limited their rights in areas such as the work- place, schools, housing, transportation systems, parks, public swimming pools, and various business establishments. Anyone breaking these laws was considered deviant and deserving of legal punishment. It took many decades of struggle, culminating in the civil rights movement of the 1950s and 1960s, to make opposing racism a socially approved rather than a deviant act. Rosa Parks, famous for her refusal to give up her bus seat to a white person, was not the only black person arrested in 1955 for violating segregation laws, but hers was the perfect "test case" to take to the Supreme Court. She was poised, respectable, and fully prepared for the negative publicity. "People always say that I didn't give up my seat because I was tired, but that isn't true," Rosa Parks (1992) wrote. "No, the only tired I was, was tired of giving in" (p. 116). Parks, an elected officer of the local chapter of the National Association for the Advancement of Colored People, had been part of the group of activists who were strategizing the best way to challenge the segregation laws, not just in court but in public awareness too. She had not planned on getting arrested that day, but when she was told to move, she recognized the opportunity. By allowing herself to be taken off the bus by the police, she provided the spark for a campaign that had already started to challenge the laws that made black Americans second-class citizens.

Parks and other activists formed the Montgomery Improvement Association, a coalition of local groups, including several church organizations, whose purpose was to demand the desegregation of buses without appearing to be too radical or threatening. A young pastor with no known enemies or personal agenda, Dr. Martin Luther King Jr., was selected to head the campaign. Despite the hardships that poor people faced, getting to work each day without public transportation, threats from motorists as they walked along the roads, and warnings that they would lose their jobs if they could not get to work each day on time, the black citizens of Montgomery organized themselves effectively and boycotted the buses for more than a year. By the time they won their case in court, they had also fostered a national reexamination of the status of black people in America. Although African Americans still face routine de facto (meaning in practice, though not in law) segregation, discrimination, and racial insults, the Montgomery bus boycott helped mobilize the tremendous effort to tear down the laws that supported and "normalized" such actions. Today it is an illegal (de jure), and thus deviant, act to deny a person equal access or equal opportunity due to his or her race.

Organizing for Environmental Change

Today, norms are being challenged to address global climate change (also known as global warming). In towns and on campuses across the United States, the norms of cranking up the heat or air conditioning and throwing away trash have been replaced with efforts to conserve energy and recycle all that citizens can. Though we have long been dependent on coal, our growing awareness of the devastating impact of coal by-products has made the "standard" black smoke–spewing coal plant a symbol of criminal negligence by corporations and the U.S. Environmental Protection Agency (EPA), which many people now view as having failed to use its full power to protect the environment. In 2008, frustrated at the inaction of the federal government, 12 states sued the EPA for violating the Clean Air Act by not setting standards to control greenhouse gas emissions in new vehicles.[1] The Supreme Court ruled that the EPA must regulate greenhouse gases. The EPA under the Obama administration agreed to do so, declaring that such emissions "threaten the public health and welfare of current and future generations" (Woodside, 2009, para. 1).

The Trump administration, however, has taken the opposite approach, minimizing and denying the science community's research and warnings regarding climate change. During his presidential bid, he declared that he would eliminate the "Department of Environment Protection" (referring to the EPA). He promised "we're going to have little tidbits left but we're going to get most of it out," and during his first two years in office, his administration reduced or eliminated nearly 80 environmental regulations, including some specifically related to coal (Feldscher, 2016; Popovich, Albeck-Ripka, & Pierre-Louis, 2017). When, in late 2018, the 1,600 page Fourth National Climate Assessment was issued by his own administration, Trump responded by stating "I don't believe it," despite overwhelming scientific evidence as to the severity of climate change and the impacts it will have on the United States and Americans in the coming years (U.S. Global Change Research Program, 2018). The report finds that climate change is having the following effects:

1. Communities
   Climate change creates new risks and exacerbates existing vulnerabilities in communities across the United States, presenting growing challenges to human health and safety, quality of life, and the rate of economic growth.

2. Economy
   Without substantial and sustained global mitigation and regional adaptation efforts, climate change is expected to cause growing losses to U.S. infrastructure and property and impede the rate of economic growth over this century.

3. Interconnected Impacts
   Climate change affects the natural, built, and social systems we rely on individually and through their connections to one another. These interconnected systems are increasingly vulnerable to cascading impacts that are often difficult to predict, threatening essential services within and beyond the nation's borders.

4. Actions to Reduce Risks

Communities, governments, and businesses are working to reduce risks from and costs associated with climate change by taking action to lower greenhouse gas emissions and implement adaptation strategies. While mitigation and adaptation efforts have expanded substantially in the last four years, they do not yet approach the scale considered necessary to avoid substantial damages to the economy, environment, and human health over the coming decades.

5. Water

The quality and quantity of water available for use by people and ecosystems across the country are being affected by climate change, increasing risks and costs to agriculture, energy production, industry, recreation, and the environment.

6. Health

Impacts from climate change on extreme weather and climate-related events, air quality, and the transmission of disease through insects and pests, food, and water increasingly threaten the health and well-being of the American people, particularly populations that are already vulnerable.

7. Indigenous Peoples

Climate change increasingly threatens indigenous communities' livelihoods, economies, health, and cultural identities by disrupting interconnected social, physical, and ecological systems.

8. Ecosystems and Ecosystem Services

Ecosystems and the benefits they provide to society are being altered by climate change, and these impacts are projected to continue. Without substantial and sustained reductions in global greenhouse gas emissions, transformative impacts on some ecosystems will occur; some coral reef and sea ice ecosystems are already experiencing such transformational changes.

9. Agriculture

Rising temperatures, extreme heat, drought, wildfire on rangelands, and heavy downpours are expected to increasingly disrupt agricultural productivity in the United States. Expected increases in challenges to livestock health, declines in crop yields and quality, and changes in extreme events in the United States and abroad threaten rural livelihoods, sustainable food security, and price stability.

10. Infrastructure

Our nation's aging and deteriorating infrastructure is further stressed by increases in heavy precipitation events, coastal flooding, heat, wildfires, and other extreme events as well as changes to average precipitation and temperature. Without adaptation, climate change will continue to degrade infrastructure performance over the rest of the century, with the potential for cascading impacts that threaten our economy, national security, essential services, and health and well-being.

11. Oceans and Coasts

Coastal communities and the ecosystems that support them are increasingly threatened by the impacts of climate change. Without significant reductions in global greenhouse gas emissions and regional adaptation measures, many coastal regions will be transformed by the latter part of this century, with impacts affecting other regions and sectors. Even in a future with lower greenhouse gas emissions, many communities are expected to suffer financial impacts as chronic high-tide flooding leads to higher costs and lower property values.

12. Tourism and Recreation

Outdoor recreation, tourist economies, and quality of life are reliant on benefits provided by our natural environment that will be degraded by the impacts of climate change in many ways.

Former U.S. vice president Al Gore and the United Nations' Intergovernmental Panel on Climate Change (IPCC) won the 2007 Nobel Peace Prize for their efforts to bring the impending environmental, social, and political catastrophes of global warming to the public's attention. The IPCC, a group of 1,300 independent scientific experts from nations all over the world maintains, in accordance with most all scientists, that human actions are fostering climate change on Earth that will likely have a dramatic impact on the planet. While climate change has occurred throughout the history of the Earth, the Industrial Revolution sparked a dramatic increase in the production of heat-trapping gases, which if left unchecked will lead to drastic environmental and social upheaval (IPCC, 2007, 2012, 2018).

The fifth assessment report by the IPCC (2013) states,

Warming of the climate system is unequivocal, and since the 1950s, many of the observed changes are unprecedented over decades to millennia. The atmosphere and ocean have warmed, the amounts of snow and ice have diminished, sea level has risen, and the concentrations of greenhouse gases have increased. (p. 3)

The atmospheric concentrations of carbon dioxide (CO_2), methane, and nitrous oxide have increased to levels unprecedented in at least the last 800,000 years. CO_2 concentrations have increased by 40% since pre-industrial times, primarily from fossil fuel emissions and secondarily from net land use change emissions. The ocean has absorbed about 30% of the emitted anthropogenic carbon dioxide, causing ocean acidification. (p. 7)

Continued emissions of greenhouse gases will cause further warming and changes in all components of the climate system. Limiting climate change will require substantial and sustained reductions of greenhouse gas emissions. (p. 14)

Other studies indicate that warmer ocean temperatures will lead to higher levels of mercury in fish, as the metabolism of fish increases in warmer waters and they eat more. Mercury in humans has been connected to medical issues such as heart attacks, high blood pressure, and learning disabilities in children (Dijkstra et al., 2013; Fears, 2013).

Most nations have outlined plans and/or already taken clear steps to reduce greenhouse gas emissions; yet in the United States, fewer than half of all Americans believe that dealing with global climate change should be a high priority for the nation, and few political leaders discuss it (Knight, Robins, Clover, & Saravanan, 2011; Pew Research Center, 2011; Rosenthal, 2011). While 97 percent of peer-reviewed articles on the topic argue that global climate change is caused, at least in part, by humans, only 48 percent of Americans recognize that fact (Funk & Kennedy, 2016). Just as there were once those who argued that tobacco was not harmful, some opponents of efforts to address global warming have tried to cast doubt on the veracity of the scientific evidence of climate change (Dash, 2011). As lethal storms and environmental disasters, like Hurricane Sandy in 2012, the typhoon in the Philippines in 2013, the massive tornadoes and flooding in parts of the United States in 2013 and 2014, again with Hurricane Harvey and Hurricane Maria in 2017 and 2018, and the flooding in Europe in 2013 and 2014 (to name just a few), accumulate, however, even many doubters are starting to realize the dangers of unchecked global warming. As the National Climate Assessment pointed out (U.S. Global Change Research Program, 2014), residents of the United States have already begun to experience its impact, and the impact will only become more severe as temperatures continue to rise. According to Oxfam, 41 million people were affected by extreme flooding between June 2017 and December 2018;

150 million people globally will live on land that is below sea level or flood level by the end of the century; and growing storm surges and tsunamis now threaten one quarter of the world's population (Oxfam, 2018).

Fortunately, efforts to address climate change exist on many fronts, including through higher education and through state, national, and global corporate initiatives. Young adults are those most likely to understand the threat of global climate change. Eighty-one percent of young people believe that global warming is real and important, and while conservatives are generally less accepting of the science, 58 percent of young conservatives say they would be less likely to vote for a candidate who opposes efforts to decrease global warming (Pew Research Center, 2018).

Despite the clear, overwhelming, and agreed upon evidence from scientists, the United States pulled out of the Paris Climate Agreement in 2017, largely considered the most important global effort to reduce climate emissions and climate change, making the United States one of the only countries not to participate. Many businesses and universities as well as 10 states, countered this move by joining the We Are Still In movement (see https://www.wearestillin.com/), pledging to continue to reduce emissions in accordance with the Paris Agreement. The United States is by far the largest contributor to global climate change, currently responsible for one third of all global carbon emissions, despite only representing 4 percent of the world's population.

Further, looking at climate change through an economic lens, the World Bank, often associated with deregulation economics, has put climate change "at the center of the bank's mission" (Davenport, 2014, para. 10).

> The World Bank Group recognizes climate change as an acute threat to global development that increases instability and contributes to poverty, fragility, and migration. Since the Paris Agreement was reached in December 2015, the World Bank Group has seen strong demand from our clients—low and middle-income countries—for rapid, concerted action on climate change. These countries recognize the threat and the opportunity: that the transition to a low-carbon, climate resilient economy can drive innovation, jobs, and growth. (World Bank, 2017)

While governments ultimately bear the responsibility to sign on to and commit to significant climate change commitments, we know that organized efforts by individuals can also have a dramatic impact on the threat of global climate change. They have already greatly reduced the threat from fluorinated gases, a type of greenhouse gas that once gravely threatened the ozone layer. These gases stem from human-made sources, such as refrigerants, fire extinguishers, and aerosol propellants (European Commission Climate Action, 2011). Thanks to the Montreal Protocol on Substances That Deplete the Ozone Layer, created in 1987, fluorinated gases have been reduced dramatically. All the nations in the United Nations signed the Montreal Protocol and agreed to reduce their use of products that emit these gases. This global effort has led to the health and survival of the ozone layer in our atmosphere and shows that global climate change can be addressed with a united global effort.

Other environmental success stories include the following:

- Since 1970, the Clean Air Act (substantially revised and expanded in 1990 and amended many times over the years) has led to a 73 percent decrease in levels of ozone, lead, carbon monoxide, nitrogen dioxide, and sulfur dioxide (EPA, 2018).

- In 2013, eight states—California, New York, Maryland, Oregon, Massachusetts, Connecticut, Rhode Island, and Vermont—promised to work together to substantially increase the number of zero-emission cars driven by ramping up the construction of charging stations. They want to make buying and charging such vehicles easy for consumers and to have 3.3 million battery-powered cars, plug-in hybrids, and other clean-burning vehicles on their roads by 2025 (Dearen, 2013).

- Thirty-eight states and the District of Columbia have Renewable Portfolio Standards that require energy providers to use renewable energy sources (e.g., solar, wind) as a portion of their energy supply (Leon, 2013).

- By 2011, the majority of nations (at least 119) had established some type of target for renewable energy or a renewable energy support policy (more than double the number that did so in 2005) (Sawin, 2011).

- In response to anticipated pressures from environmentalists and to help their own bottom line, "Western multinationals—and in some cases, their Asian suppliers—have in the last five years started to build more environmentally sound factories in developing countries, green-building experts say" (Ives, 2014, para. 5).

- In the United States and Canada, many hundreds of colleges and universities have made effective efforts to increase recycling efforts and reduce their greenhouse gas emissions and use of nonrenewable energy. In collaboration with the U.S. Green Building Council, *The Princeton Review* (2018) publishes

The Princeton Review's Guide to 339 Green Colleges, which describes how these schools have displayed "a strong commitment to sustainability in their academic offerings, campus infrastructure, activities and career preparation."

- Most of the world came together to agree to the Paris Climate Agreement in 2015, a task that many thought was unlikely if not impossible, requiring countries to reduce emissions according to their pledges.

- In January 2019, 65 countries ratified the Kigali Amendment to the Montreal Protocol, pledging to reduce hydrofluorocarbons (HFCs) by more than 80 percent over the next 30 years. As more countries ratify this, it is estimated that it can reduce an increase of .7 degrees of global warming (UN Environment, 2019).

These efforts show signs of some progress and the potential for more. However, for us to effectively mitigate and adapt to global climate change (it is too late to stop it), more people must become aware of its reality and potential impact. Without a more educated public, we cannot achieve the united global effort needed to deal with this serious global threat.

Sociologist in Action

Ellis Jones

On April 22, 1990, something changed for me. It was a Sunday. I was a student at the University of Southern California, and a friend invited me to walk down to a fair that was going on nearby. I didn't have any plans, so I decided to join him. Apparently, it was something called "Earth Day." There were tables and booths everywhere, and people were excitedly milling about from one to the next. Every booth I visited offered me a different way to make a positive environmental impact: recycling, composting, conserving water, reusing old clothes, . . . the list went on and on.

At the end of that day, I felt absolutely inspired to make the world a better place. In my mind, the environmental movement had experienced a stroke of true genius. They weren't asking people to join groups, attend meetings, or organize rallies. Instead, they had opened up a completely new realm of action for

(Continued)

people—their own everyday lives. By engaging people in micro activism rather than asking them to commit to the much more intensive work undertaken by full-time activists, they were essentially democratizing access to social change. It was a way to expand the environmental movement to almost anyone despite their limits of time, money, skills, or circumstances.

As soon as I realized this, I imagined that thousands of others were having the same revelation at that same moment of history. The environmental movement was laying the foundation for all other social movements to do exactly the same thing! The human rights movement, the social justice movement, the animal rights movement, the feminist movement, the LGBT (lesbian, gay, bisexual, and transgender) movement, and all the rest would begin to adapt this new technique to engage people everywhere in actions they could take in their own lives to contribute to some much-needed social change.

Granted, no particular action would add up too much in terms of impact, but if enough people were inspired to act, the collective impact of thousands or tens of thousands or perhaps even millions of people would redefine what kind of changes were possible around these issues. At the same time, all of these people would now feel more invested in the outcomes that they themselves had been working toward in their own way. Rather than relying solely on a relatively small group of activists to bear the weight of these daunting social and environmental problems, now everyone could take on a piece of that responsibility.

The more I thought about this potential, the more I became convinced that what people really needed (or at least what I really needed) was a book that would collect as many of these actions as possible into a single place. I was absolutely sure that someone would write it and I, in turn, would be first in line to buy it. Ten years later, I was a sociology graduate student at the University of Colorado, Boulder. I had waited, patiently, for the imaginary author of this hypothetical book to appear, and she or he had yet to step forward. In the meantime, I was teaching Social Problems, along with a number of other mainstay sociology courses, and found that my own students were constantly asking what they could actually do about all of these problems we were spending so much time reading about, discussing, and analyzing.

I had waited 10 years and decided that was long enough. So I roped in two of my closest friends (also fellow sociology grad students), and we spent the next year collecting all of the actions we could find, from every source we could get our hands on, and distilled the results into a single book. We added a section summarizing the latest data on some of the most significant social and environmental problems we seemed to be facing at the beginning of the 21st century, and the result was *The Better World Handbook: Small Changes That Make a Big Difference*. It has since sold 25,000 copies, been added to more than 300 college and university libraries worldwide, and been adopted in sociology classrooms across the country. But perhaps more important, it's even more widely used by people who've never taken a sociology class in their life.

The website (betterworldhandbook.com) has had more than 550,000 unique visitors since it was created. In a very practical sense, sociology provided the three of us with the tools we needed to uncover, understand, and translate our world's social and environmental problems into a form that allows each of us to personally contribute to their resolution.

Source: Courtesy of Ellis Jones.

Exercise 6.1
Nonverbal Sanctions

Next time you see someone give a nonverbal, negative sanction (a look of disgust, a shake of the head, or some other nonverbal sign of disapproval), ask yourself the questions below.

As you answer these questions, be sure to define norms and socialization and distinguish between folkways and mores.

1. What norm was being broken (i.e., what was the act that led the person to give a nonverbal negative sanction)?

2. What was the reaction of the norm violator to the negative sanction?

3. What was your reaction to the interaction you witnessed between the norm violator and the person giving the negative sanction?

4. Why do you think you reacted this way? How did your own socialization process influence your reaction?

Exercise 6.2
Breaching Folkways

Often, we are so used to following the norms of society, we forget that we could behave differently. Harold Garfinkel (1967), a sociologist most famous for his "breaching" experiments, in which people would break folkways to expose them, had students do things such as act like boarders in their parents' home, offer to pay store clerks less or more money than the price indicated on an item, and question every statement a person made in a conversation. Through these experiments, his students challenged the norms of social interaction and, in the process, brought them to light.

Conduct your own breaching experiment on campus by carrying out the following steps:

1. The next time someone asks you how you are doing after saying hello, tell the person how you are really doing. In fact, take several minutes to thoroughly relate how you are feeling at that time.

2. Repeat Step 1 with four different people.

3. In a one- to two-page paper, describe (a) how the four people reacted to your experiment, (b) how you felt as you carried out the experiment, (c) how their and your reactions related to your relationship with them, and (d) what you learned about norms of interacting through carrying out this experiment.

Exercise 6.3
Sexual Assault and Sexual Harassment

1. Read "A New Survey Finds 81 Percent of Women Have Experienced Sexual Harassment" (https://www.npr.org/sections/thetwo-way/2018/02/21/587671849/a-new-survey-finds-eighty-percent-of-women-have-experienced-sexual-harassment).

2. Write a one-page paper outlining the information from the article and offering your analysis.

3. Next, go to RAINN's (Rape, Abuse, and Incest National Network) State Law Database (https://apps.rainn.org/policy/). Then (a) enter the zip code of your home town; (b) enter the zip code of your university (if you attend university in a state different from where your home town is located); (c) enter the zip code from a state whose laws you suspect will be quite different from those of your state. With each, read through the variety of state laws regarding sexual assault, rape, and sex crimes.

4. Write a one- to two-page paper analyzing this information. What did you learn about some of the laws that are in place? Were some laws stronger than others? Did some states have more or stronger laws than others? What does this tell you about the values, mores, and norms of each state on this issue? Of society in general?

Exercise 6.4
Global Climate Change and Pollution

One of the major social problems we all face today is global climate change. Another, some might say, is that many of us don't believe we have the power to make a difference on policy levels on issues as big and important as climate change.

1. Read "No Vote, No Problem. These Young Environmental Activists Think Outside the Ballot Box" at https://www.motherjones.com/environment/2018/10/no-vote-no-problem-these-young-environmental-activists-think-outside-the-ballot-box/ and "Meet Young Activists of Color Who Want to Combat Climate Change and Save the Planet" (https://www.usatoday.com/story/news/politics/onpolitics/2018/08/20/zero-hour-young-people-color-climate-change-awareness/797489002/).

2. Then listen to rap "Broken," featuring Xiuhtezcatl Martinez (https://www.youtube.com/watch?v=LKUZJjxm9Vs).

(Continued)

(Continued)

3. Write a two-page paper reflecting on the two articles and the video. What did you learn? How did it affect you? What does it tell you about the power of (young) people to make a difference?

4. Create a one- to two-page action plan outlining how you could affect policy change on climate change either on your campus or in your community.

5. Share your plan with three other classmates, and together combine your plans to create one synthesized, two-page action plan to present to your class and that could work to affect change in your community.

Exercise 6.5
What Is Your Ecological Footprint?

Go to the website http://www.myfootprint.org and determine your ecological footprint and your own impact on global climate change.

1. Describe your ecological footprint.

2. Were you surprised by what you found? Why or why not?

3. Go to Earthwatch at www.earthwatch.org, click on your part of the world and read about some of the ways you can make a difference, and then go to UN Environment at https://www.unenvironment.org/get-involved/campaigns and read about some of their campaigns to create social change. How might you work with others to use social capital to connect (a) different groups on your campus, (b) different groups in your surrounding community, and (c) your campus and surrounding community to work together to address global climate change?

Exercise 6.6

Google and Duke University Team Up to Capture Methane and Show How It Can Be Done!

1. Learn about the Google and Duke University Partnership to Capture Methane at http://seedstock.com/?s=duke+and+google+hog+waste, and read the *Duke Carbon Offsets Initiative Annual Report* (https://sustainability.duke.edu/sites/default/files/DCOI%20annual%20report_2017.pdf).

2. Then, write a two-page essay that describes what you learned from the report and the article and how they influenced your perspective on global climate change and how it should be tackled. Be sure to include a discussion of the roles of governments, corporations, and higher education in addressing climate change.

Exercise 6.7

Lunch Counter Sit-Ins and Other Nonviolent Protests

One of the more dramatic forms of protest during the civil rights movement was the "lunch counter sit-in." At these events, blacks took seats at lunch counters that served only whites and then refused to leave. During many of these sit-ins, white customers jeered, threw food at them, and threatened them, before the police came and dragged off the protesters.

Use your library's electronic databases (e.g., the JSTOR or Academic Search Premier database) to find three academic articles that focus on lunch counter sit-ins during the civil rights movement. Also, watch the first segment of the *Eyes on the Prize* documentaries (your school library should have

(Continued)

(Continued)

a copy or you can find it online), in which the sit-ins are covered. Then, write a two- to three-page paper that addresses the following questions:

1. Given the fact that the protesters knew that they were not going to be served, what did they hope to gain by doing this? Were they trying to change the law? Public opinion? The policies of the lunch counter establishments?

2. Do you think their strategy was successful? What are the benefits and positives of nonviolent protest? What are the negatives and limitations?

3. Considering public opinion toward black Americans today and given what you have learned about social movements, (a) what sorts of public actions could black Americans do today to bring attention to the kinds of discrimination they face and (b) what actions can you personally take that would help diminish racial discrimination? Hand in a complete bibliography of your sources along with your written answers.

Exercise 6.8
Studying a Social Movement

There have been numerous social movements (organized campaigns for social, cultural, or legal change) in the United States that have met with varying degrees of success over the past 50 years. The cultural period known as "the sixties" (roughly 1963–1973) saw the rise of the second wave of the women's movement, the American Indian Movement, the Brown Power movement (Latinos), an antiwar movement (Vietnam), an international student and youth movement, and movements for poverty relief and fair housing practices, gay rights, environmental rights, and animal rights, among others.

Pick one of these movements. Use your library's databases to search for articles using key words and phrases such as "American Indian Movement," "organization, successes, failures, media portrayal," and "goals." Then, answer the following questions:

1. What were or are the goals of the movement you have chosen?

2. How was or is the movement organized? How was or is it run? How was or is it viewed in the popular media? How do you think the movement was or is organized, and how has its portrayal in the media affected its ability to achieve its goals?

3. How successful or unsuccessful has the movement been in achieving its goals?

4. Imagine you were hired as a consultant to promote and advise the movement. What would you do or say to help it achieve its goals? Be specific and strategic, outlining plans to carry forth your proposal.

5. Extra credit: Identify a current social movement (Me Too; Black Lives Matter; March for Our Lives, etc.). Follow the steps of the above exercise in examining one of these movements. Additionally, how are today's movements similar to and/or different from past movements? Are there parts of today's movements you find particularly effective? Can you see yourself participating in one? Why or why not?

Exercise 6.9
Campus Activism: Then and Now

The 1960s were famous for student activism and student organizing. Who were these students, and for what were they advocating? How were they portrayed in the news media at the time? How are they being portrayed by the media and history textbooks today? How successful were

(Continued)

(Continued)

they? Write a three- to four-page paper that answers these questions and describes your reaction to what you have read by carrying out the following steps:

1. Go to the newspaper archive in your library.

2. Choose one regional or national paper for which there are issues available from 1963 through 1973.

3. Scan through the issues published between 1963 and 1973 until you have located (and photocopied) at least 10 articles about on-campus student activism.

4. For each article, note the following: (a) Does the article tell you what the students wanted? (b) Does the article tell you why they had those goals? (c) Does the article appear to have a position in support of, neutral to, or opposed to the students? (d) For each case in which you have a clear sense of the students' goals, were they achieved? Did they change society?

5. Can you imagine 1960s-style student protests happening today on your campus? If yes, would the protesters have similar or different goals from those of the 1960s protesters you read about? If no, explain why you think they would not take place today. Have other styles of protest replaced them? Have students become less likely to protest and, if so, why?

6. Only *after* you have answered Question 5, research one of the following movements: Occupy Wall Street; Fair Trade Campaigns; student-led campus living wage campaigns; Carry That Weight; Black Lives Matters; March for Our Lives; or campus fossil fuel divestment campaigns. In a one- to two-page paper, outline (a) what you have learned about this movement; (b) how does learning about this movement change or add to your analysis in Questions 1 through 5?

Exercise 6.10
The Disability Rights Movement

Read "Welcoming Art Lovers With Disabilities" at https://www.nytimes
.com/2013/10/27/arts/artsspecial/welcoming-art-lovers-with-disabilities
.html and view, listen to, and read the Disability Rights Movement display at http://www.americanhistory.si.edu/disabilityrights/exhibit.html and
read the article "These High Schoolers Are Calling for a National Disability
History Museum by Making Their Own" (https://www.wbur.org/hereand
now/2018/07/04/national-disability-history-museum-high-schoolers).
Then, answer the following questions:

1. According to the Smithsonian's Disability Rights Movement
   display, what do those in the movement seek to gain? How have
   some of their tactics mirrored those of other social movements?

2. Have you ever acted to support the efforts of the Disability Rights
   Movement? Why or why not?

3. Describe how you could help achieve the goals of the Disability
   Rights Movement. How do the actions of the high school students
   from Gann Academy inform your thinking?

4. How would the success of the Disability Rights Movement benefit
   all members of society, whether or not they are individuals with
   disabilities?

Exercise 6.11
Organizing to Address Hunger

1. Check out the videos and other sources of information on the Take
   Part website at http://www.takepart.com/place-at-the-table/index.html
   and read about the different demographic groups facing hunger on

(Continued)

(Continued)

the Feeding America website at http://feedingamerica.org/hunger-in-america.aspx to create a flyer about hunger that you can pass out in the cafeteria at your school.

2. Bring the flyers to your school cafeteria at a mealtime.

3. Sit down with different groups of students you do not know, give them your flyer, and engage them in a conversation about hunger in the United States. Then, ask them if they would like to work with you to find a way to assist the people in your area who are hungry. Be sure to put your contact information on the flyer and to have a sign-up sheet so that those who are interested can give you their names and contact information.

4. Set a meeting time, and send an e-mail out to announce the meeting. Then, meet with those interested, and during the meeting, formulate a plan to help the hungry in your area. For tips on how to run a meeting, you can look at page 4 of this website from WE Charity (formerly Free the Children) (http://assets.filemobile.com/15/get-involved/resources/Campus%20in%20Action%20Toolkit.pdf).

5. Once you have formulated your plan, carry it out.

6. Report on (a) how you organized your group, (b) the results of your group's efforts, and (c) what students in general can do to help address the issue of hunger (based on what you learned on the hunger websites) in a letter to the editor or an article in your school's newspaper.

DISCUSSION QUESTIONS

1. What was the last norm that you broke? Think of the last folkway (rather than a more) that you violated. What was the reaction of those around you? How did it make you feel? Did the reaction affect your willingness to break the folkway in the future?

2. Imagine you own a global corporation. Why should you care about global climate change? What steps will you take to protect your business

from its harmful impacts? How does your perspective on global climate change when you imagine you are an owner of a global corporation differ from the one you may usually have? Why?

3. Have you ever broken a rule or law that you considered unjust? Why did you do so? Did you organize your "deviant" behavior with other people, or did you violate the norm on your own? What were the repercussions? Did your action result in any lasting social change? Why or why not?

4. What are the norms that guide your classroom behavior? What would you consider deviant classroom behavior by students? How did you learn these norms? What are the negative sanctions for those who violate classroom norms?

5. Twenty-three percent of women and 5.5 percent of men experience rape or sexual assault while in college. Why do you think rates are so high on campus? Is there anything about the culture, values, and norms of your campus that might contribute to this? As an individual, what can you do, and what can you do in organizing a group of students to help change this culture?

6. Are there any laws that you find to be unfair or legal punishments for crimes committed that you think are too strong, harsh, or outdated? If not, why not? If so, what are they? Why do you think these laws and punishments exist even though they are unfair? What values (and whose values) do they reflect? What do these laws and punishments teach you about power relations in U.S. society?

7. Think of some examples that support Durkheim's idea that deviance can serve positive functions for society. Now think of how deviance can have a negative influence on society. What are the situations in which deviance is most likely to have (a) a positive or (b) a negative effect on society?

8. Do you conserve and recycle as much as you can? If so, why? If not, what would convince you to do so?

9. Review Durkheim's description of organic solidarity, and then describe why it is in the self-interest of most global business leaders to address global climate change. Why is it in your self-interest?

10. Recall what Cooley meant by the term looking-glass self. Do you think your college classmates see you as the type of person who would (a) participate in or (b) lead a protest on your college campus? Do you think your high school classmates would give the same answers? Why or why not? If you consider your generalized other (Mead), is it possible to change this?

SUGGESTIONS FOR SPECIFIC ACTIONS

1. Go to the Teaching Tolerance website at https://www.tolerance.org/professional-development/strategies-for-reducing-racial-and-ethnic-prejudice-essential-principles. Use the 13 principles outlined there to organize a group action around diversity and equity on your campus. Be sure to obtain your instructor's approval before beginning your project. Actions may be local or national, educational, or political. Keep a journal of your project, its outcome, the feedback you get from others, and what you learned by doing it.

2. Identify a social issue that you support. (Consider movements to promote fair trade products, establish equal rights for LGBTQ Americans, create universal health care, etc.) Find a campus or local organization that works on this issue.

   a. Attend meetings, and participate in the activities of the organization.

   b. Write a letter to the campus newspaper explaining to your classmates what the issues are that the group is concerned with, how these issues affect students, how the group is trying to change people's opinions and feelings on this issue, and why they should join the organization too!

   Please go to this book's website at http://study.sagepub.com/white6e to find further civic engagement opportunities, resources, peer-reviewed articles, and updated web links related to this chapter.

NOTE

1. You can read the states' petition (http://www.pewclimate.org/docUploads/Mass-v-EPA-Petition.pdf).

REFERENCES

Becker, H. (1963). *Outsiders: Studies in the sociology of deviance*. New York, NY: Free Press.

Calma, J. (2018, Oct. 1). No vote, no problem. These young environmental activists think outside the ballot box. *Mother Jones*. Retrieved from https://www.motherjones.com/environment/2018/10/no-vote-no-problem-these-young-environmental-activists-think-outside-the-ballot-box/

Carlsen, A., Salam, M., Miller, C. C., Lu, D., Ngu, A., Patel, J. K., & Wichter, Z. (2018, October 23). #MeToo brought down 201 powerful men. Nearly half of

their replacements are women. *The New York Times*. Retrieved from https://www
.nytimes.com/interactive/2018/10/23/us/metoo-replacements.html

Chatterjee, R. (2018, February 21). A new survey finds 81 percent of women have experienced sexual harassment. NPR.org. Retrieved from https://www.npr
.org/sections/thetwo-way/2018/02/21/587671849/a-new-survey-finds-eighty-percent-of-women-have-experienced-sexual-harassment

Dash, J. (2011, December 6). *The Contrarians: Full essay with references*. Retrieved from http://climate.uu-uno.org/articles/view/148487/?topic=50699#Quacks

Davenport, C. (2014, January 24). Industry awakens to threat of climate change. *The New York Times*. Retrieved from http://www.nytimes.com/2014/01/24/science/earth/threat-to-bottom-line-spurs-action-on-climate.html

Dearen, J. (2013, October 25). 8 states join to promote clean cars. *The Boston Globe*. Retrieved from http://www.bostonglobe.com/news/nation/2013/10/24/states-join-forces-promote-clean-cars/4aTkMlnkleI9rhYwGCMcGO/story
.html?s_campaign=8315

Dijkstra, J. A., Buckman, K. L., Ward, D., Evans, D. W., Dionne, M., & Chen, C. Y. (2013, March 12). Experimental and natural warming elevates mercury concentrations in estuarine fish. *PLoS ONE*. Retrieved from http://www.plosone.org/article/info%3Adoi%2F10.1371%2Fjournal.pone.0058401

Duke University. (2017). *Duke carbon offset initiative 2017 annual report*. Retrieved from https://sustainability.duke.edu/sites/default/files/DCOI%20annual%20
report_2017.pdf

Durkheim, É. (1982). *The rules of the sociological method* (S. Lukes, Ed., W. D. Halls, Trans.). New York, NY: Free Press. (Original work published 1895)

Elignon, J. (2012, March 22). Taking on police tactic, critics hit racial divide. *The New York Times*. Retrieved from http://www.nytimes.com/2012/03/23/nyregion/fighting-stop-and-frisk-tactic-but-hitting-racial-divide.html

Environmental Protection Agency. (2018). *Detailed summary: Clean Air Act results*. Retrieved from http://www.epa.gov/air/caa/progress.html#pollution

European Commission Climate Action. (2011, September 26). *Fluorinated greenhouse gases*. Retrieved from http://ec.europa.eu/clima/policies/f-gas/index_en.htm

Fears, D. (2013, October 13). Study links warmer water temperatures to greater levels of mercury in fish. *The Washington Post*. Retrieved from http://
www.washington post.com/national/health-science/study-links-warmer-water-temperatures-to-greater-levels-of-mercury-in-fish/2013/10/13/c86d43c6-3113-11e3-9c68- 1cf643210300_story.html

Feldscher, K. (2016). Trump says he'd eliminate "Department of Environment Protection." *Washington Examiner*. Retrieved from https://www.washingtonexaminer
.com/trump-says-hed-eliminate-department-of-environment-protection

Funk, C., & Kennedy, B. (2016). *Public views on climate change and climate scientists*. Pew Research Center. Retrieved from http://www.pewinternet.org/2016/10/04/public-views-on-climate-change-and-climate-scientists/

Garfinkel, H. (1967). *Studies in ethnomethodology*. Englewood Cliffs, NJ: Prentice Hall.

Goldstein, J. (2013, August 14). Police dept's focus on race is at core of ruling against stop-and-frisk tactic. *The New York Times*. Retrieved from http://www.nytimes.com/2013/08/15/nyregion/racial-focus-by-police-is-at-core-of-judges-stop-and-frisk-ruling.html?hp

Intergovernmental Panel on Climate Change. (2007). *Climate change 2007: Synthesis report*. Retrieved from http://www.ipcc.ch/publications_and_data/ar4/syr/en/spm.html

Intergovernmental Panel on Climate Change. (2012). *Managing the risks of extreme events and disasters to advance climate change adaptation* (C. B. Field, V. Barros, T. F. Stocker, D. Qin, D. J. Dokken, K. L. Ebi, . . . P. M. Midgley, Eds.; A special report of Working Groups I and II of the Intergovernmental Panel on Climate Change). Cambridge, England: Cambridge University Press. Retrieved from http://www.ipcc.ch/pdf/special-reports/srex/SREX_Full_Report.pdf

Intergovernmental Panel on Climate Change. (2013, September 27). *Working Group I contribution to the IPCC fifth assessment report climate change 2013: The physical science basis summary for policymakers*. Retrieved from http://graphics8.nytimes.com/packages/pdf/science/27climate-ipcc-report-summary.pdf

Intergovernmental Panel on Climate Change. (2018). Global warming of 1.5 degrees celsius. Retrieved from https://www.ipcc.ch/sr15/

Ives, M. (2014). Slowly, Asia's Factories Begin to Turn Green. *The New York Times*. Retrieved from https://www.nytimes.com/2014/01/08/business/international/asian-factories-see-sense-and-savings-in-environmental-certification.html

Knight, Z., Robins, N., Clover, R., & Saravanan, D. (2011, January 13). *Climate investment update* (HSBC Global Research). Retrieved from http://www.research.hsbc.com/midas/Res/RDV?ao=20&key=68W7FaNQA0&n=288184.PDF

Lamberth, J. (1994). *Revised statistical analysis of the incidence of police stops and arrests of black drivers/travelers on the New Jersey Turnpike between Interchanges 1 and 3 from the years 1988 through 1991* (Unpublished manuscript). Department of Psychology, Temple University, Philadelphia, PA. Retrieved from http://www.mass.gov/eopss/docs/eops/faip/new-jersey-study-report.pdf

Leon, W. (2013, June). *The state of state renewable portfolio standards*. Retrieved from http://www.cleanenergystates.org/assets/2013-Files/RPS/State-of-State-RPSs-Report-Final-June-2013.pdf

Livingston, G., & Brown, A. (2017). *Intermarriage in the US 50 years after Loving v. Virginia*. Pew Research Center Report.

Mohn, T. (2013, October 25). Welcoming art lover with disabilities. *New York Times*. Retrieved from https://www.nytimes.com/2013/10/27/arts/artsspecial/welcoming-art-lovers-with-disabilities.html

National Sexual Violence Resource Center. (NSVRC) (2019). *Get statistics: Sexual assault statistics*. Retrieved from https://www.nsvrc.org/statistics.

New Jersey Division of Criminal Justice. (2009). *Eradicating racial profiling companion guide*. Retrieved from http://www.state.nj.us/lps/dcj/agguide/directives/racial-profiling/pdfs/ripcompanion-guide.pdf

New Jersey Division of Criminal Justice. (n.d.). *Overview of New Jersey's racial profiling policy*. Retrieved from http://www.state.nj.us/lps/dcj/agguide/directives/racial-profiling/pdfs/overview-racial-policy.pdf

Parks, R. (with Haskins, J.). (1992). *Rosa Parks: My story*. New York, NY: Dial Books.

Pew Research Center. (2011, June 10). *Views of Middle East unchanged by recent events*. Retrieved from http://www.people-press.org/2011/06/10/views-of-middle-east-unchanged-by-recent-events

Pew Research Center. (2018). *The generation gap in American politics*. Retrieved from http://www.people-press.org/2018/03/01/the-generation-gap-in-american-politics/

Popovich, N., Albeck-Ripka, L., & Pierre-Louis, K. (2017, October 05). *78 environmental rules on the way out under Trump*. Retrieved from https://www.nytimes.com/interactive/2017/10/05/climate/trump-environment-rules-reversed.html

The Princeton Review. (2018). *The Princeton Review's guide to 339 green colleges*. Retrieved from http://www.princetonreview.com/green-guide.aspx

Rape, Abuse & Incest National Network. (n.d.). *Campus sexual violence: Statistics*. Retrieved from https://www.rainn.org/statistics/campus-sexual-violence

Reiman, J., & Leighton, P. (2017). *The rich get richer and the poor get prison: Ideology, class, and criminal justice* (10th ed.). Boston, MA: Allyn & Bacon.

Rosenthal, E. (2011, October 15). Where did global warming go? *The New York Times*. Retrieved from http://www.nytimes.com/2011/10/16/sunday-review/whatever-happened-to-global-warming.html?pagewanted=all

Sawin, J. L. (2011). *Renewables 2011 global status report*. Retrieved from http://german watch.org/klima/gsr2011.pdf

Smith, S. G., Basile, K. C., Gilbert, L. K., Merrick, M. T., Patel, N., Walling, M., & Jain, A. (2017). *National intimate partner and sexual violence survey (NISVS): 2010-2012 state report*. Atlanta, GA: National Center for Injury Prevention and Control, Centers for Disease Control and Prevention.

Sutherland, E. H. (1947). *Principles of criminology* (4th ed.). Philadelphia, PA: J. B. Lippincott.

UN Environment. (2019, January 3). World takes a stand against powerful greenhouse gases with implementation of Kigali Amendment [Press release]. Retrieved from https://www.unenvironment.org/news-and-stories/press-release/world-takes-stand-against-powerful-greenhouse-gases-implementation

U.S. Global Change Research Program. (2014). *National Climate Assessment*. Retrieved from http://nca2014.globalchange.gov/

U.S. Global Change Research Program. (2018). Fourth National Climate Assessment. Retrieved from U.S. Global Change website at https://nca2018.global change.gov/

Woodside, C. (2009, May 19). *Was an "historic" EPA ruling on GHGs reflected by historically good coverage?* (The Yale Forum on Climate Change and the Media). Retrieved from http://www.yaleclimatemediaforum.org/2009/05/epa-ruling-on-ghgs/

World Bank Group. (2017, June 1). World Bank Group statement: Our commitment to action on climate change [Press release]. Retrieved from http://www.worldbank.org/en/news/press-release/2017/06/01/world-bank-group-statement-our-commitment-to-action-on-climate-change

7

Big Money Doesn't *Always* Win

Stratification and Social Class

*Money, money, money . . . it's
a rich man's world.*[1]

—ABBA

D o you agree with the lyrics of this ABBA song? Do you think "it's a rich man's world"? In this chapter, we will discuss social stratification—how societies distribute the things of value to them and rank groups of people according to their access to what is valued. Although people may prize certain things more in some societies than in others (e.g., privacy, tea or coffee, certain spices, etc.), almost everyone everywhere places a high value on money.

In the United States, the desire for money, combined with a lack of government oversight of some banking practices, led to the Great Recession of 2008 to 2009, from which we have not yet fully recovered. The Glass-Steagall Act, passed in 1933 after the stock market crash of 1929, separated commercial banking (which accepts deposits and lends money) from investment banking (which issues securities and invests using credit). These changes were enacted to prevent banks from taking investment risks that could jeopardize their solvency. However, banks, seeking the ability to make more money, lobbied for the repeal of the act. Once the Glass-Steagall Act was repealed in 1999, commercial banks were once again able to invest their clients' deposited money and practice both investment and commercial banking within one company.

During the same time period, under a drive to give all Americans the opportunity to own their own homes, government mortgage providers Fannie Mae and Freddie Mac lowered their standards and began issuing mortgages to more people with below-average credit ratings and low incomes (Roberts, 2008). Meanwhile, as government regulators turned a blind eye, other mortgage lenders followed suit; housing prices rose dramatically as more and more people bought homes, and it became a seller's market. With no preventative government regulations, the mortgage lenders profited enormously by collecting fees to guide homeowners into taking on mortgages they could not afford. The lenders bundled toxic mortgages (whose owners were likely to default) into "collateralized debt obligations" (CDOs). They then divided the bundles into more CDOs and sold them to banks, which sold them to investors, which purchased insurance against possible losses on the bundled mortgages. The lack of government regulation and adequate oversight allowed the companies that insured the CDOs to fail to maintain the resources needed to pay possible insurance claims. Therefore, when the homeowners defaulted on loans, they were never able to afford and investors filed claims to cover their losses on CDOs, the insurance companies were unable to cover the losses, and the entire system fell apart.

Credit rating agencies, whose job it is to grant credit scores that provide a rating of securities, also failed to do their jobs responsibly. These credit rating agencies (Moody's, Standard and Poor's, and Fitch), paid by the banks to which they issue credit scores, granted them high credit ratings, leading investors in these banks into thinking that their money was not at risk (Morgenson, 2008). When these CDOs started to fail, many investors lost money, and some banks, such as Bear Stearns and Lehman Brothers, went bankrupt, while others were saved through government bailouts to prevent further damage to the economy (The Financial Crisis Inquiry Commission, 2011).

As a result of the above events, the stock market suffered the worst decline since the Great Depression, and the housing market collapsed, as the worth of homes plummeted and toxic mortgages and high unemployment rates led to defaults on mortgages. Retirees lost money they counted on for living expenses. Institutions that relied on interest and dividends from investments (e.g., state pension funds, charities, colleges, art museums, and grant funders) faced drastic cuts in their income, and many had to stop providing services and lay off employees. As is generally the case, the most vulnerable workers were hit the hardest by the recession. The working poor have faced the most job cuts and the greatest number of lost work hours. As Andrew M. Sum, director of the Center for Labor Market Studies at Northeastern University, states, "Low-income people are the big losers when the economy turns down" (quoted in Eckholm, 2008, para. 6). A loss of a few hours of work each week, never mind a layoff, can be devastating for workers already trying to scrape by on low-wage work.

While the U.S. economy recovered well during the Obama administration, its growth has been uneven, and inequality has increased (see Figure 7.1). The economy has recovered for the very wealthy; however the average family income, when controlled for inflation, is actually lower today than it was 25 years ago (DeNavas-Walt, Proctor, & Smith, 2013; Irwin, 2013). Today, the richest 1 percent of U.S. households owns 40 percent of all U.S. wealth, more wealth than *all* of the bottom 90 percent combined (Wolff, 2017)! Income and wealth inequality has not been so uneven since before the Great Depression hit in 1929. The top 1 percent of Americans also made 26.3 times as much money (income) as the bottom 99 percent of

Figure 7.1 An Uneven Recovery

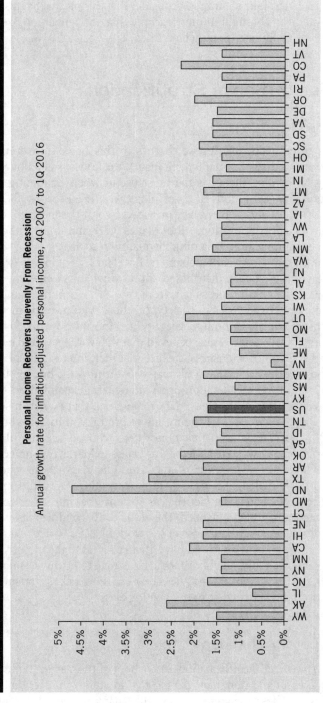

Personal Income Recovers Unevenly From Recession

Annual growth rate for inflation-adjusted personal income, 4Q 2007 to 1Q 2016

Source: Data from the Pew Charitable Trusts analysis of the U.S. Bureau of Economic Analysis' State Quarterly Personal Income and Annual State Personal Income, accessed March 26, 2019.

Note: Pew included the District of Columbia when calculating personal income growth for the U.S. However, Pew did not break out results for the District.

Americans in 2015. Take a minute to consider that this means that just over 3 million Americans brought in more income than the remaining 324 million Americans combined (Sommeiller, Price, & Wazeter, 2016).

Theories of Social Stratification

Classical Theories

Of the three main theories we discuss in this book (functionalism, conflict theory, and symbolic interactionism), the consensus among sociologists is that conflict theory is the most useful when examining social stratification. The two founders of sociology who used the conflict perspective, Karl Marx and Max Weber, provided two of the most important classical theories of stratification. All of Marx's work and much of Weber's focused on examining and explaining inequality in society.

As you will recall from earlier chapters, Karl Marx spent his life examining how power is unequally distributed and how we might change society to make this distribution process fair. In most of his work, he maintained that there are only two classes in society: (a) those who own the means of production (the owners or bourgeoisie) and (b) those who work for them (the workers or proletariat). Marx believed that the workers would eventually develop a "class consciousness," realize that their interests are in opposition to those of the owners, and ultimately overthrow the bourgeoisie.

Weber, who lived a generation or so after Marx, expanded on Marx's ideas (see Weber, 1968). Whereas Marx maintained that power in society directly relates to ownership of the means of production (those who own industries), Weber recognized that nonowners who possess useful skills can also have some power. He added a third class to Marx's two-tiered class system. He divided the nonowners into a middle class (those who had skills based on knowledge) and a working class (those who did manual labor). Weber anticipated the fact that as societies became more complicated and technologically advanced, the skills of the middle class would become more in demand than those of the working class. The middle class would, therefore, be paid more and have greater access to the things valued in society (better schools, neighborhoods, housing, etc.) than the members of the working class. Unlike Marx, he didn't envision a class uprising that would result in the destruction of the owning class.

Contemporary Theories

Today, *power elite theories*, a modern offshoot of the classic conflict perspective, represent the dominant perspective in studies of economic and political power. These theories maintain that many people do not bother participating in the political process because they feel alienated from it. Power elite theorists, such as C. W. Mills (1956/1970) and William Domhoff (1967/2005, 1983), note that a relatively small, organized group of people hold key positions in the major institutions of society and make continuous (and, overall, successful) efforts toward keeping themselves in power. Mills noted a connection among key players in the military, corporations (that supply the military and help finance campaigns for those seeking office in government), and the government (which chooses which corporations will supply the military). Domhoff (1967/2005) goes further, arguing that there is a "ruling class" composed of interrelated leaders in the corporate, political, and policy-planning network worlds. Members of this group grew up together, attending the same private schools, social clubs, and colleges. As adults, they occupy positions of power and "rule" the United States.

Mills's focus on the military and military industrial complex as part of the ruling political elite is significant. Today, U.S. military spending is $585 billion, occupying a significant portion of the aggregate U.S. budget, an increase of 31 percent from $384 billion in 2000 (White, 2017; Arthur and Frisk, 2014). Out of each dollar Americans pay in taxes, 37 cents are allocated to Pentagon spending for past and present wars, while only 19 cents is spent on health care and 15 cents responding to poverty (White, 2017, 209–210). In his chapter "A Gluttonous Military Budget Leaves Our Social Welfare in Poor Health," in the book *Preventing War and Promoting Peace: A Guide for Health Professionals* by William Wiist and Shelley White, Jonathan examines the military budget even further:

- To bring every person in the United States who lives below the poverty line (15%; 47 million people) to above the poverty line, would cost $175 billion. The United States could pay this amount for the next 30 years for the amount spent to date for the wars in Afghanistan and Iraq ($46 trillion).

- Sixteen million U.S. children, or 22%, live below the poverty line. Children therefore comprise 32% of all poor people in the United States. The Children's Defense Fund estimates that by expanding investments in existing policies and programs that work, we can shrink overall child poverty 60 percent, Black child poverty 72 percent, and improve economic circumstances for 97 percent of poor children at a cost of $77.2 billion a year. This could be paid for with a 50 percent reduction in the 800 military bases the United States operates worldwide.

- In 2014, there were approximately 4.5 million children under the age of 19 in the United States without health insurance. Using one year of the DOD budget, we could ensure all kids are insured for the next 48 years or equally we could do so through an annual $12 billion reduction to the DOD budget.

Bringing the conflict perspective to a global level, the sociologist Charles Derber (2000, 2003) describes a world in which corporations rather than nations increasingly dominate. Derber maintains that governments act disproportionately in the interest of global corporations rather than their own citizens. He calls for a global democratic grassroots movement, such as the Occupy movements, to change the mode of globalization and put "people before profit" (Derber, 2003). In doing so, he claims that we can reduce inequalities between social classes across the globe.

Social Class

What social class are you? Wait, we can guess (with at least 95 percent certainty) how you will answer! When asked on surveys, almost all Americans will say that they are middle class. Most, however, have no social scientific understanding of what "social class" actually means.

Neither do most Americans have any idea of the level of economic inequality in the United States. In one recent study, a representative sample of Americans were asked to look at three charts that represented different distributions of wealth and to guess which one indicated the wealth distribution in the United States. The chart selected by the majority of respondents actually showed the distribution of wealth in Sweden—the nation with the *greatest* economic equality! When they were asked which would be the most ideal degree of wealth distribution in a society, they selected the chart that indicated an even *more* equitable distribution of wealth than that found in Sweden. Few people were aware of the lopsided distribution of wealth in the United States and that the richest 20 percent of Americans actually own approximately 84 percent of the wealth, compared with 58 percent for the top 20 percent in Sweden and 32 percent for the wealthiest 20 percent in the "ideal" nation (Norton & Ariely, 2011).

We need to recognize economic inequality in order to understand the influence of social class on our society and on the lives of individual people. According to social scientists (those who measure class professionally), people in the same social class have relatively equal access to what is valued in their society (e.g., money, power, good schools, nice neighborhoods, etc.) and have similar lifestyles. Once people realize this, they can usually recognize the many differences concealed by the almost universal "middle class" label in the United States. For example, it quickly becomes clear that whereas both high-level managers and sales associates tend to refer to themselves as middle class, members of these respective occupations do not have similar lifestyles or access to what is valued in U.S. society. And all this doesn't even address the fact that your answer to the question "What social class are you?" was likely an answer about your perceived class *in the United States* and not about your perception of your class within the global economic system. In the global economic system, the majority of Americans would be considered upper class.

Because social scientists understand that almost all Americans will simply put down "middle class" if asked to specify their social class, most of them measure social class by looking at the income, education, and profession of the respondents. However, this still leaves some room for error.[2] In general, most truly middle-class people in the United States have some college education, have professions with salaries (rather than jobs where they are paid by the hour), and earn within a specified range around the median income. Sometimes, social scientists break down members of the middle class into two groups: middle class (white-collar workers) and working class (blue- or pink-collar workers[3]). Blue- and pink-collar jobs, which usually do not require as much education, tend to pay less than white-collar work.[4] White-collar work tends to require at least a bachelor's degree.

The importance of attaining a college degree cannot be overstated. Those who hope to attain a class ranking of at least middle class generally need to have one or more of the following: (a) a high level of education, (b) connections to those in power, and (c) a legacy of wealth passed down to them by their family. While one has to be born lucky to have the third requirement (passed-down wealth), the second (connections to those in power) also requires a certain degree of good fortune. You need to have access to people in power in order to establish connections with them. A high degree of education, while increasingly expensive and open only to those with a certain level of aptitude, remains the most readily available path to a middle-class life for most Americans. The gaps in income levels and employment rates between those with and without a college degree have never been greater. In 2017, the unemployment rate for those with a four-year college degree was 2.5 percent, compared with 3.4 percent for those with a two-year degree and 4.6 percent for those with only a high school diploma (BLS, 2018b). Likewise, only 4.5 percent of those with four-year college degrees were in poverty, compared with 9.4 percent of those who had completed some college, 13.3 percent of those whose had earned a high school diploma but not attended college, and 24.8 percent of those who did not graduate high school (Semega, Fontenot, & Kollar, 2017).

Sex and race also relate to social class. In short, it helps to be male and white. Female heads of households are almost twice as likely as male heads of families and five times as likely as married couples to live below the poverty level. In 2017, an estimated 12.3 percent of Americans lived in poverty. The poverty rate also varies dramatically according to race. In 2017, 18.3 percent of Hispanics, 21.2 percent of blacks, 10 percent of Asians, and 26.2 percent of Native Americans lived in poverty (Fontenot, Semega, & Kollar, 2018; Statista, 2018; Wilson & Mokhiber, 2017). We'll talk more about the connections between sex and race and social class in Chapters 8 and 9 of this book.

Social Class and Political Representation

Voter turnout is strongly related to social class. Poor and working-class people are much less likely to vote than middle-class and wealthier Americans. Although always a feature of U.S. elections, the class gap has now become a chasm. Many poor Americans do not see the issues that they care about being addressed by politicians and, therefore, see no use in going through the hassles of registering to vote and voting. Some are also working so hard, piecing together two or three jobs, and so preoccupied with finding a way to pay the rent and buy food that they have little time or energy to vote. Homeless Americans, without proof of residence, face even more difficulty in exercising their right to vote. There also is a gap between poor and non-poor Americans on which candidates they support. In the 2016 presidential election, the bottom two quintiles, those earning under $30,000 and $30,000 through $49,999 respectively, voted for Hilary Clinton by approximately 10 point margins. The other three quintiles, with incomes ranging from $50,000 to $250,000 voted for Donald Trump, by much slimmer margins ranging from 1 to 4 percent (Huang, Jacoby, Strickland, & Lai, 2016). If voters from the bottom two quintiles had turned out in larger numbers, they may likely have changed the result of the election.

Education affects voter turnout as well. Voter rates by education level in the 2016 election illustrate this with a 35 percent voting rate for those with less than a high school education, 51 percent for those with a high school diploma, 69 percent for those with some college education or a college degree, and 80 percent for those with a graduate degree (United States Election Project, n.d.).

The decline of union power has particularly diminished the influence of working-class Americans on the electoral process. According to the Bureau of Labor Statistics, union membership in 2017 had dropped to half of what it was in 1983, with a 34.4 percent membership rate for public sector workers and 6.5 percent for private sector workers (BLS, 2018a). Wages of union workers remained much higher than those of nonunionized workers.

In 2017, the median weekly salary of full-time, nonunion workers was $829, while the median weekly salary of full-time union workers was $1,041, leaving them with just 80 percent of the income of their unionized counterparts (BLS, 2018a). In addition to ensuring relatively good wages, unions have traditionally been a very powerful "get out the vote" force for (usually Democratic) candidates. However, as the power of unions has declined, so has the ability of the working class to have its economic interests addressed by politicians.

Even Americans who do vote realize that only a very small percentage of the population has the means to run for high political office. The amount of money people must be able to raise to establish a legitimate campaign prohibits the vast majority of people from running for such positions. For example, during the 2016 congressional elections, the average candidate winning a position in the House of Representatives or the Senate spent $1,300,000 and $10,600,000, respectively (Kim, 2016). At the presidential level, the numbers are even more staggering: Hilary Clinton, Donald Trump, and their allies raised over a billion dollars combined during the presidential election of 2016 (Blumenthal, 2016).

The overall result of the current electoral system[5] in the United States is that most poor and working-class Americans neither vote nor run for political office. In turn, they are largely ignored by political officeholders, who must respond to constituents who do vote and contributors who help them raise the vast sums of money needed to gain and maintain their positions.

The Power of Organized People

The information above may sound depressing to anyone committed to true democracy in the United States. However, it is only part of the picture. Although civic organizations may not exist in the size and number they once did (Skocpol, 2003), there are many grassroots political organizations successfully representing lower income Americans. While we usually do not hear about their efforts on the evening news, Barack Obama's experience as a community organizer who later became President helped more people become aware of their existence. These organizations are accomplishing the vital task of giving voice to poor and working-class persons.

These civic groups, in essence, teach people to use a "sociological eye" and a "sociological imagination" as they train them to become effective citizens. They train ordinary Americans to question the status quo, connect their personal problems to public issues, and hold politicians and business leaders accountable to the citizenry. While there are many of these organizations, a common principle is that they all organize for power. Their ability to organize allows them to act as a kind of mediating institution for the poor, working, and middle classes and to negotiate effectively with those who have political and financial power.

The following description of UnidosUS/National Council of la Raza, located on the organization's website (UnidosUS, n.d.), provides an insight into the efforts of these types of organizations:

> We serve the Hispanic community through our research, policy
> analysis, and state and national advocacy efforts, as well as in our
> program work in communities nationwide. And we partner with
> a national network of nearly 300 Affiliates across the country to
> serve millions of Latinos in the areas of civic engagement, civil
> rights and immigration, education, workforce and the economy,
> health, and housing.

South Bronx Churches (SBC), an Industrial Areas Foundation (IAF) affiliate composed of 8 Catholic and 17 Protestant churches, also provide a good example of the power of organized people. During the 1970s and early 1980s, the South Bronx community in New York City was suffering. Largely abandoned by government and business leaders, drugs were everywhere, shootings were commonplace, and arson was incessant. A turning point came when SBC members began to talk to one another about their fears, frustrations, and hopes concerning their neighborhoods. They realized that alone, none of them could do much, but together they had the power to improve their community. Part of their multifaceted plan to revitalize the area was the "Nehemiah Homes" project, the creation of 1,000 houses for poor, working-class residents.[6] SBC's effort to reinvigorate the South Bronx was a long struggle that included public confrontations with powerful mayors of New York City (David Dinkins, Ed Koch, and Rudolph Giuliani) and huge efforts to raise money from business leaders. In the end, the SBC won. The organization succeeded in raising millions of dollars and convincing the city of New York to donate land for the project. Today, 1,000 low-income, working New Yorkers own their homes where there were once just burned-out and decrepit buildings.

The National Employment Law Project (NELP) provides another example of a successful organizing effort. NELP has effectively brought together existing organizations throughout the United States to create fairer minimum wage laws with their "Raise the Minimum Wage" campaign. As of June 2018, 29 states had raised their state minimum wage levels above that of the federal $7.25 minimum, and several more were expected to do so soon. NELP helped establish a broad-based movement to harness the will of 72 percent of the U.S. public (including 88 percent of Democrats, 75 percent of Independents, and 48 percent of Republicans) and a broad coalition of community organizing groups in favor of raising the minimum wage and has effectively pushed for such legislation in states and many cities across the nation (Bloomberg, 2014; NELP, 2015).

Other organizations—such as the National Urban League, which has a mission "to enable African Americans and other underserved urban residents to secure economic self-reliance, parity, power and civil rights;" the Children's Defense Fund, a "child advocacy organization that has worked relentlessly for more than 40 years to ensure a level playing field for all children;" the Center for Budget and Policy Priorities, which pursues "federal and state policies designed both to reduce poverty and inequality and to restore fiscal responsibility in equitable and effective ways. We apply our deep expertise in budget and tax issues and in programs and policies that help low-income people, in order to help inform debates and achieve better policy outcomes;" the National Coalition for the Homeless, "a national network of people who are currently experiencing or who have experienced homelessness, activists and advocates, community-based and faith-based service providers, and others committed to a single mission: "to prevent and end homelessness while ensuring the immediate needs of those experiencing homelessness are met and their civil rights protected;" the Center for Community Change, which has a mission "to build the power and capacity of low-income people, especially low-income people of color, to change their communities and public policies for the better;" the National Housing Law Project, which works to advance housing justice for poor people and communities "by strengthening and enforcing the rights of tenants, increasing housing opportunities for underserved communities, and preserving and expanding the nation's supply of safe and affordable homes;" and many others—work on the frontlines of working to change poverty and related issues in the United States. (The websites for these organizations, respectively, are http://nul.iamempowered.com/, https://www.childrensdefense.org/, https://www.cbpp.org/, https://national homeless.org/, https://communitychange.org/, https://www.nhlp.org/)

Sociologist in Action

Frances Fox Piven

Frances Fox Piven is a sociologist who never ceases to use her sociological research to influence society. When awarded the 2003 American Sociological Association's Award for the Public Understanding of Sociology, the distinguished professor of political science and sociology at the Graduate School and University Center of the City University of New York was described as "a scholar who is equally at home in the university setting and the world of politics" (ASA, 2003, para. 25). The author or coauthor of renowned sociological texts on the disenfranchisement and political power of poor Americans (e.g., *Regulating the Poor*, 1972/1993; *Poor People's Movements*, 1977; *The New Class War*, 1982/1985; *Why Americans Don't Vote*, 1988; *Why Americans Still Don't Vote*, 2000), she puts her knowledge into action in the political arena. For example, in the 1960s, she used her research to expand welfare benefits. Her efforts to enfranchise the poor were instrumental in establishing the National Voter Registration Act of 1993 (popularly known as the Motor Voter Act).[7] She has fought an unrelenting battle against the "welfare reform" initiated in 1996 and has been a consistent proponent of the politics of disruption and mass protest.

(Continued)

(Continued)

In one article, Piven (2005) describes her advice for democratic reform movements today in the following way:

> Yes, we should work on our agenda of democratic reforms, including a national right to vote, a national voter registration system, the implementation of the National Voter Registration Act, Election Day a holiday, nonpartisan election officials, and so on. But we have to do more. . . . The time when mass protest is possible will come. We should be ready and receptive, obdurate and bold. The hip-hop voter registration campaign had a slogan, "Vote or die." They were on the right track. (para. 7)

Piven is one of the boldest living sociologists. Importantly, she is a dedicated social scientist as well as an activist. She conducts good social scientific research that can be examined and critiqued objectively.

Although not everyone or even all sociologists would agree with Piven's strategies for political action, many have been inspired by her efforts to make our society a democracy in which people of all classes are represented equally.

Sociologist in Action

Adrian Chevraux-Fitzhugh

Did you know that corporations have the legal right, under the First Amendment, as interpreted by the Supreme Court's decision in *Citizens United v. Federal Election Commission* (2010), to initiate and participate in political campaigns? Adrian Chevraux-Fitzhugh used the sociological tools he gained as a sociology master's student at Humboldt State University to examine people's knowledge about "corporate personhood" and their attitudes toward corporate involvement in political elections. He designed a countywide telephone survey for Democracy Unlimited of Humboldt County (DUHC, 2005), which revealed that

> only 35% of respondents were familiar with the concept of "corporate personhood," which is the legal doctrine that allows corporations to

claim the same constitutional rights of human beings. . . . [Moreover], a significant majority opposes allowing corporations to participate in local elections. (para. 4)

These results were utilized by DUHC in its ongoing efforts to educate citizens about the relationship between corporations and democracy in the United States. DUHC helped establish the organization Move to Amend, a coalition aiming to amend the Constitution to abolish corporate personhood (https://movetoamend .org). If you'd like to find out more about corporate personhood and determine whether you think it makes sense in a democracy, please go to www.pbs.org/now/ politics/corprights.html.

Exercise 7.1
Pressures Facing the U.S. and World Economies Today

Go to *The New York Times'* "Economic Crisis and Market Upheavals" (http:// topics.nytimes.com/top/reference/timestopics/subjects/c/credit_crisis/index .html?scp=1-spot&sq=credit%20crisis&st=cse).

Read three of the articles under "Latest Developments," and relate their contents to the information in this chapter. Be sure to describe how the information from the articles helped you better understand the economic pressures on the world and U.S. economies today.

Exercise 7.2
How Unequal Is Wealth Distribution in the United States—and Why?

Look at the video on wealth inequality at (https://www.youtube.com/watch? time_continue=10&v=QPKKQnijnsM).

Now, write a one- to two-page essay that (a) describes your reaction to the video and (b) explains the information in it, using information from this chapter (and other parts of the book, if you like).

Exercise 7.3
Social Class and You

At this point in the chapter, you are probably thinking about your own social class and how you became a member of it. Think about how the institutions of family, education, and occupation are related to one another, and answer the following questions:

1. What is the highest level of education completed by each of your parents?

2. Were you raised by two parents? If not, who raised you?

3. When you were growing up, what was the occupation of the head(s) of your household?

4. Have you been in different social classes at different times in your life? If so, why? If not, why not?

5. How do you think your answers to Questions 1 to 3 influenced your present social class?

Exercise 7.4
Food Banks and Hunger in America

Volunteer to serve a meal at an area soup kitchen. Your Campus Activities Office should be able to help you and even connect you with a group on campus that regularly volunteers at one.

1. Set up a time when the manager can give you some background on why the guests come to the soup kitchen. (In particular, ask what makes them need the services the soup kitchen offers.)

2. When you are there, be sure to pay attention to the people eating at the soup kitchen (while still taking time to interact with them, to serve them, and to enjoy your volunteer experience), and note the following: (a) Are they mostly families or individuals?

(b) What is their racial and ethnic makeup? (c) What is the age range of those who have come to eat there? (d) Do most of them seem hopeful? Resigned? Angry? Happy? Depressed? (e) How do they react to your being there, and what are the ways they interact with you?

3. Go to the website of Feeding America at http://feedingamerica .org, click on "Hunger in America," and then read the Hunger Fact Sheets on this page. Incorporate the information from these fact sheets into a two- to three-page paper that describes (a) the services the soup kitchen offers, (b) the demographics and attitudes of the people eating at the soup kitchen (based on the data you have gathered in Step 2), (c) why the people who eat there need to do so, (d) whether you could ever see yourself needing the services of a soup kitchen (and why or why not), and (e) how your experience at the soup kitchen made you feel.

Exercise 7.5
America: Who Votes?

1. Click on the latest edition of the webinar "America Goes to the Polls" (http://www.nonprofitvote.org/voter-turnout.html).

2. Compare voting rates by race, income, and educational attainment.

3. Based on the information in this chapter, why do you think the respective groups are more or less likely to vote than other Americans? What does this tell you about the connections (at least those that are perceived) between wealth and power?

4. What steps do you think should be taken toward engaging nonvoting Americans enough so that they will register and vote?

5. Read "They Don't Really Want Us to Vote" at https://www.nytimes. com/2018/11/03/us/politics/voting-suppression-elections.html and "Voting Laws Roundup 2017" (https://www.brennancenter.org/analysis/ voting-laws-roundup-2017).

(Continued)

(Continued)

6. Did you know that there are many laws aimed at discouraging voter registration? What does this tell you about democracy in 2019? Why do so many of these laws target poor people, people of color, immigrants, and other disempowered demographic groups?

7. Write a report with a five-point strategic plan to eliminate voter suppression from the next elections. Compose the report as if you will be presenting it to either your state legislature or the federal legislature.

Exercise 7.6
The Influence of Money on State Politics

It's almost impossible to be unaware of the influence of money on presidential politics. However, money influences politics at the state level as well. Go to http://www.followthemoney.org and submit your address. After looking through the findings for your area, answer the following questions:

1. What are three things you learned about the interaction between politics and money in your area from this website?

2. Were you surprised by any of your findings? Why or why not?

3. How can you use the information you found to influence campaign finance laws?

Exercise 7.7
Community Organizing

Find out what community-based organizing associations are operating in your area. You can do this by contacting your campus community center and asking for referrals to local community organizing groups or by checking out the websites of national organizations (Children's Defense Fund, Community Change, Public Citizen, AFL-CIO, ACLU, etc.) and finding the respective affiliates in your state.

Attend a meeting of a local community organization and write a three-page paper that discusses (a) the issues on which the organization is currently focusing, (b) how it decided on those issues, and (c) the organization's strategy as it works on those issues. Discuss how these three items relate to (a) the sociological eye and (b) the sociological imagination and what suggestions you can make that might help the organization to be even more successful and powerful in accomplishing its mission. Finally, include a section analyzing how your university (or students from your university) could support the work of the organization.

Exercise 7.8
Fair Trade on Your Campus

1. Watch http://www.youtube.com/watch?v=NhQJrz-aDfI&feature=r elated; then, go to https://fairtradecampaigns.org/campaign-type/ universities/ and find two other sources on your own that describe the Fair Trade movement.

2. Write a two-page paper in which you use the sources above to describe the Fair Trade movement today, and describe why your campus should or should not be a part of it (you can do this even if your campus is a Fair Trade University).

3. In addition, include a one-page strategy plan for working with your school cafeteria to provide fair trade coffee and other fair trade alternatives for students or for working with your school bookstore to offer fair trade (or sweat-free) clothing and food options (or to expand these offerings if some are already available on your campus).

Exercise 7.9
Voting Rights

Go to "Voting Rights (Registration and Requirements)" (https://www.nytimes .com/topic/subject/voting-rights-registration-and-requirements).

(Continued)

(Continued)

Read three to five of the articles that focus on the efforts to promote either voter ID laws or the Voting Rights Act, and then, in two to three pages

1. summarize the information in the articles and

2. explain whether or not you support (a) voter ID laws or (b) the Supreme Court decision to overturn key portions of the Voting Rights Act.

Be sure to back up your arguments with hard evidence (and cite your sources).

Exercise 7.10
Power and the Control of Seeds

Today, three corporations control more than half of all the commercial seed sold throughout the world (Center for Food Safety, 2013). Learn about how this trend relates to power elite theories on class conflict by completing the following exercises.

Watch "Seeds of Freedom" at http://topdocumentaryfilms.com/seeds-of-freedom, and then watch the brief video "Reclaiming Our Hearts and Home" at the Navdanya website (http://www.navdanya.org).

Now read "Seeds and Genetic Diversity" at http://www.etcgroup.org/issues/seeds-genetic-diversity, and look up the information on the Just Label It website (http://justlabelit.org).

In one to two pages, answer the following question:

*   Describe how a power elite theorist like Charles Derber might explain the centralization of food production under the control of huge companies like Monsanto.

Exercise 7.11
Corporations Are People?

1. Watch the video "The Story of Citizens United v. FEC" at https://storyofstuff.org/movies/story-of-citizens-united-v-fec/ and read "Timeline on Corporate Personhood" (https://www.citizen.org/sites/default/files/students-citizens-united-guide-corporate-personhood.pdf).

2. What did you learn? How are corporations legally defined as people and what is the history that led to this? What are some of the concerns raised in the video regarding this and what are some of your own concerns?

3. Create a fact sheet or pamphlet to educate your campus (and beyond) on the issue of corporate personhood, including some resources of organizations working to challenge it.

DISCUSSION QUESTIONS

1. Before reading this chapter, did you know what a large percentage of the U.S. budget is spent on defense spending? Were you surprised when reading about some of the "trade-offs," ways in which reducing military spending and reassigning that money could have a great effect on poverty reduction? Go to https://www.costofwar.com and spend at least five minutes looking at and understanding the time tickers in the middle of the page. What are your thoughts after spending time viewing these?

2. Why do you think so many people who are not part of the 1 percent are opposed to raising taxes on the wealthiest members of society? How would Marx explain their thinking (recall his concept of "false consciousness")? Do you agree with Marx? Why or why not?

3. Reread the discussion of Durkheim's ideas on external and internal inequality in Chapter 2 of this book. How might external inequality in the United States today influence who might be elected to public office?

4. Think of an issue relating to social class inequality you'd like to see addressed, and conduct a power analysis of who controls the decisions about that issue and how you might influence them. For example, say you want your school to serve fair trade coffee (if it does not already). Who decides what coffee the school purchases to sell on campus? How might you convince him or her to switch to fair trade coffee?

5. Have you ever worked (or are you now working) at a relatively low-wage hourly job? Why or why not? Do you think your wage is fair? Why or why or not?

6. What is your immediate reaction when you think about the fact that one in four households with children reports an inability to buy enough food? Why? How can more people become informed about this social problem and take action to alleviate it?

7. Do you think unions are needed for low-wage workers? Why or why not?

8. What do you think your life would be like if you worked at a low-wage hourly job for the rest of your life? How would it affect whom you may marry, what kind of family you will have, what kind of home you will live in, what you will do for entertainment, and so forth?

9. How, if at all, do you think the college degree you are planning to get will affect your social class? Why?

10. How did your social class influence your decision to (a) attend college and (b) attend the particular college in which you are enrolled?

11. Why might it make good economic sense for corporations to "go green"? If you were a CEO of a major corporation, what might prompt you to invest funds in reducing your company's "environmental footprint"? Do you think more companies will do so soon? If so, which ones (and where and why)?

SUGGESTIONS FOR SPECIFIC ACTIONS

1. Find a local community organizing association in your area. (You can do this by looking for affiliated organizations at the websites of the Community Change, Children's Defense Fund, Public Citizen, AFL-CIO, ACLU, and so forth.) Contact the lead organizer and set up a meeting with him or her. Find out what issues the organization is working on at this time. Offer to use your skills as a college student to conduct some basic research for the group to help it learn about the issues.

2. Go to the Rock the Vote website at https://www.rockthevote.org/ and follow the directions there on how to conduct a voter registration drive on your campus.

3. Go to Inequality.Org's website at https://inequality.org/ and click on "Action." Look through the list of actions they are working on and participate in one or organize a teach-in at your home or school on a current issue related to social class, stratification, and the economic recession.

4. Go to the Food Research and Action Center's (FRAC) website at http://frac.org/ and go the "Legislative Action Center." Read about the various legislative actions this organization is undertaking, and choose one to participate in. Create a campaign to educate your campus community about the legislative item and to encourage them to join you in taking action on it. Please go to the book's website at http://study.sagepub.com/white6e to find further civic engagement opportunities, resources, peer-reviewed articles, and updated web links related to this chapter.

NOTES

1. The lyrics of "Money, Money, Money" by ABBA were retrieved from http://www.metrolyrics.com/money-money-money-lyrics-abba.html
2. For instance, this doesn't take into account student loans, whether one has a spouse who is working outside the home, whether one has children to support, and so forth.
3. White-collar workers are office workers, including managers, professionals, and other educated and salaried workers. Blue-collar work refers to manual unskilled or semiskilled labor, such as that carried out by mechanics, plumbers, and factory workers. Pink-collar work is unskilled or semiskilled work traditionally carried out by women, such as work done by waitresses, clerks, secretaries, and florists.
4. In general, skilled laborers earn higher wages than unskilled blue-collar workers. Some skilled blue-collar workers even earn higher salaries than some white-collar workers. For example, in 2004, the median salary of mental health workers (white collar) was $36,630, and the median salary for electricians (blue collar) was $45,200.
5. Actually, we do not have one unified electoral system. Elections are handled at the county and state level, with rules for registering voters, designing ballots, the type of voting machine, and so forth, determined by local officials rather than by the federal government.
6. Anyone who knows anything about the tremendous need for affordable housing in New York City (and across the nation) will particularly appreciate this story.
7. This act was a compromise that led to allowing those applying for a driver's license to register to vote. Poor people applying for government aid programs

were also supposed to be encouraged to register. Unfortunately, although many, many Americans register to vote while acquiring their driver's license, relatively few government officials have followed through with the Motor Voter Act's requirement that they register poor people.

REFERENCES

Anker, R., & Anker, M. (2017). *Living wages around the world: Manual for measurement*. Northampton, MA: Edward Elgar Publishing.

Arthur, D., & Frisk, D. (2014, November). *Growth in DoD's budget from 2000 to 2014*. Washington, DC: Congressional Budget Office.

ASA. (2003). ASA award recipients honored in Atlanta. *Footnotes*. Retrieved from http://www.asanet.org/sites/default/files/savvy/footnotes/septoct03/fn4.html

Bloomberg. (2014). *U.S. issues*. Retrieved from http://www.bloomberg.com/info graphics/2014-03-12/national-poll.html#us-issues

Blumenthal, P. (2016, September 21). Hillary Clinton continues to build campaign money advantage over Donald Trump. Retrieved from https://www.huffingtonpost .com/entry/2016-campaign-money_us_57e2163ee4b0e28b2b511a7f

Brennan Center for Justice. (2017). Voting laws roundup 2017. Retrieved from https://www.brennancenter.org/analysis/voting-laws-roundup-2017

Bureau of Labor Statistics. (2018a January 19).Union membership (annual) news release [Press release]. Retrieved from https://www.bls.gov/news.release/union2 .htm

Bureau of Labor Statistics. (2018b, March 27).Unemployment rates and earnings by educational attainment. Retrieved from https://www.bls.gov/emp/chart-unemployment-earnings-education.htm

Center for Food Safety. (2013, February 13). *Seed giants vs. U.S. farmers*. Retrieved from http://www.centerforfoodsafety.org/reports/1770/seed-giants-vs-us-farmers

Citizens United v. Federal Election Commission, 558 U.S. 310 (more) 130 S. Ct. 876 (2010).

Democracy Is for People.org. (n.d.). Timeline on corporate personhood. Retrieved from https://www.citizen.org/sites/default/files/students-citizens-united-guide-corporate-personhood.pdf

Democracy Unlimited of Humboldt County. (2005, March 30). *Survey says: Yes to local democracy, no to corporate campaign contributions*. Retrieved from http://www.ufppc.org/us-a-world-news-mainmenu-35/2704-news-northern-california-county-survey-shows-only-35-have-heard-of-corporate-personhood.html

DeNavas-Walt, C., Proctor, B. D., & Smith, J. C. (2013, September). *Income, poverty, and health insurance coverage in 2012* (U.S. Census Bureau). Retrieved from http://www.census.gov/prod/2013pubs/p60-245.pdf

Derber, C. (2000). *Corporation nation: How corporations are taking over our lives—and what we can do about it.* New York, NY: St. Martin's Griffin.

Derber, C. (2003). *People before profit: The new globalization in an age of terror, big money, and economic crisis.* New York, NY: Macmillan.

Domhoff, G. W. (1983). *Who rules America now?* New York, NY: Simon & Schuster.

Domhoff, G. W. (2005). *Who rules America? Power, politics, and social change* (5th ed.). New York, NY: McGraw-Hill. (Original work published 1967)

Eckholm, E. (2008). Working poor and young hit hard in downturn. *The New York Times.* Retrieved from http://www.nytimes.com/2008/11/09/us/09young.html?scp=1&sq=working%20poor%20and%20recession&st=cse

The Financial Crisis Inquiry Commission. (2011). *The financial crisis inquiry report.* Retrieved from http://www.gpo.gov/fdsys/pkg/GPO-FCIC/pdf/GPO-FCIC.pdf

Fontenot, K., Semega, J., & Kollar, M. (2018). Income and poverty in the United States: 2017. *United States Census Bureau, 14.*

Huang, J., Jacoby, S., Strickland, M., & Lai, K. R. (2016, November 9). Election 2016: Exit polls. *The New York Times.* Retrieved from https://www.nytimes.com/interactive/2016/11/08/us/politics/election-exit-polls.html

Irwin, N. (2013, September 17). The typical American family makes less than it did in 1989. *The Washington Post.* Retrieved from http://www.washingtonpost.com/blogs/wonkblog/wp/2013/09/17/the-typical-american-family-makes-less-than-it-did-in-1989/

Kim, S. R. (2016, November 9). The price of winning just got higher, especially in the Senate. Retrieved from https://www.opensecrets.org/news/2016/11/the-price-of-winning-just-got-higher-especially-in-the-senate/

Mills, C. W. (1970). *The power elite.* New York, NY: Oxford University Press. (Original work published 1956)

Morgenson, G. (2008, October 22). Credit rating agency heads grilled by lawmakers. *The New York Times.* Retrieved from http://www.nytimes.com/2008/10/23/business/economy/23rating.html?_r=1

National Employment Law Project. (2015, May). *Minimum wage basics: Public opinion on raising the minimum wage* [Fact sheet]. Retrieved from https://www.nelp.org/wp-content/uploads/Minimum-Wage-Basics-Polling.pdf

Norton, M., & Ariely, D. (2011). Building a better America—one quintile at a time. *Perspectives on Psychological Science, 6*(1), 9–12.

Pew Research Center. (2016, July 28). Personal income growth shows uneven U.S. recovery. Retrieved from https://www.pewtrusts.org/en/research-and-analysis/articles/2016/07/28/personal-income-growth-shows-uneven-us-recovery

Piven, F. F. (2005, Winter). Voting and voters. *Logos.* Retrieved from http://www.logosjournal.com/issue_4.1/piven.htm

Roberts, R. (2008, October 3). How government stoked the mania. *The Wall Street Journal*. Retrieved from http://online.wsj.com/article/SB122298982558700341.html

Semega, J. L., Fontenot, K. R., & Kollar, M. A. (2017). Income and poverty in the United States: 2016. *United States Census Bureau*, 10–11.

Skocpol, T. (2003). *Diminished democracy: From membership to management in American civic life*. Norman, OK: University of Oklahoma Press.

Sommeiller, E., Price, M., & Wazeter, E. (2016). Income inequality in the US by state, metropolitan area, and county. *Economic Policy Institute*.

Statista. (2018, September). U.S. poverty rate in 2017, by ethnic group. Retrieved from https://www.statista.com/statistics/200476/us-poverty-rate-by-ethnic-group/

Story of Stuff Project. (n.d.). *Story of Citizens United v. FEC*. Retrieved from https://storyofstuff.org/movies/story-of-citizens-united-v-fec/

UnidosUS. (n.d.). We are Unidos US. Retrieved from https://www.unidosus.org/about-us/

United States Election Project. (n.d.). Voter turnout demographics. Retrieved from http://www.electproject.org/home/voter-turnout/demographics

Weber, M. (1968). *Economy and society*. Totowa, NJ: Bedminster Press.

White, J. (2017). A gluttonous military budget leaves our social welfare in poor health. In W. Wiist & S. White (Eds.), *Preventing war and promoting peace: A guide for health professionals* (pp. 205–216). Cambridge, England: Cambridge University Press. doi:10.1017/9781316551158.018

Wilson, V., & Mokhiber, Z. (2017, September 15). 2016 ACS shows stubbornly high Native American poverty and different degrees of economic well-being for Asian ethnic groups. [Blog post]. Retrieved from https://www.epi.org/blog/2016-acs-shows-stubbornly-high-native-american-poverty-and-different-degrees-of-economic-well-being-for-asian-ethnic-groups/

Wolff, E. N. (2017). *Household wealth trends in the United States, 1962 to 2016: Has middle class wealth recovered?* (No. w24085). Cambridge, MA: National Bureau of Economic Research.

CHAPTER 8

What Does a "Typical American" Look Like Today?

Race and Ethnicity

D oes race matter? How you answer that question has a lot to do with your own experiences and your knowledge about society. Your answer also depends on your understanding of how racial groups have been treated throughout the history of the United States. In this chapter, we will examine the relationships among immigration and race and ethnicity,[1] the social construction of race, and the persistence of racism in the "color-blind" era.

Immigration and Race/Ethnicity

Emblazoned on the Statue of Liberty is a poem by Emma Lazarus (1883/2009) representing the statue as the "mother of exiles" as it proclaims "worldwide welcome" to those spurned by other nations. The mother of exiles tells the other nations to

Give me your tired, your poor,

Your huddled masses yearning to breathe free,

The wretched refuse of your teeming shore.

Send these, the homeless, tempest-tost to me.

Ironically, this poem was written a year after the enactment of the first law to restrict immigration in the United States, the Chinese Exclusion Act of 1882.

Whereas earlier immigrants came primarily from northern and western Europe, the late 1800s saw southern and eastern Europeans seeking refuge from persecution or economic opportunity. On the West Coast of the United States, Chinese and Japanese immigrants also began to arrive, looking for jobs. Pseudoscience that "proved" that western Europeans were a race superior to eastern Europeans, southern Europeans, and all people of color spurred negative *racial prejudice* (irrational feelings about a racial group) and *racial discrimination* (actions for or against people based on their race) against the groups deemed to be inferior. Racism, which is a system of advantage based on concepts of race (Wellman, 1993), can be observed through institutional policies, such as immigration. Racism, coupled with periodic economic downturns that led to increased competition for work, resulted in restrictive, race-based immigration policies that remained in place until 1965.

After the Chinese Exclusion Acts of 1882 and 1892, and the Gentlemen's Agreement of 1907, which largely curtailed Chinese and Japanese immigration respectively, a further series of laws prohibited or limited immigration of non-western Europeans. These restrictions on immigration culminated in the National Origins Acts of 1921 and 1924, which established quotas (by nation of origin) and allowed only a trickle of immigration to continue. The 1924 Immigration Act enacted 2 percent immigration quotas per nation, based on the 1890 U.S. Census.[2] So, for example, Italy could send only 2 percent of the number of Italian people residing in the United States in 1890. Since there were mostly whites in the United States at the time, from various European countries, this ensured that the nation would grow as a predominantly white nation, with little immigration allowed from Asian, Latin American, and African nations. A provision to the 1924 act limited immigration to those eligible for citizenship. Because only those of white and African American descent could become citizens, this, in effect, prohibited all further Asian immigration.

These restrictions remained in effect until the Immigration Act of 1965 was passed at the height of the civil rights movement and amid worldwide pressure to overturn legalized racial discrimination in the United States. It abolished national quotas (replacing them with quotas for the Eastern and Western Hemispheres) and did much to increase immigration and alter the racial makeup of the United States. The foreign-born population rose from 4.7 percent in 1970 (Gibson & Lennon, 1999) to 13.7 percent in 2017 (Lange & Torbati, 2018). It is important to understand that people do not generally leave their homes and families unless they have very strong reasons to do so. Immigrants move to new nations due to a variety of "push" and "pull" factors. Push factors can include lack of economic opportunity in their home countries, war, famine, genocide, political persecution, and so on. Pull factors from the receiving nation can include economic expansion and better opportunities for jobs, peace, political and religious freedoms, and family or friends who have already emigrated. If you look at the history of immigration to the United States, you can see the impact of these factors on immigration policies (relatively open when we are in peaceful, economic boom times and much more restrictive when we are at war and/or in economic downturns).

A change in push and pull factors has resulted in the recent downturn in immigration to the United States from Mexico from 11.7 million per year in 2009 to 11.3 million per year in 2018 (Henderson, 2018). Anti-immigration measures have increased in the United States amid persistently high unemployment since the Great Recession of 2008 to 2009 and significantly under the Trump administration and Republican controlled Congress of 2016 through 2017 (Pierce & Selee, 2017). Meanwhile, Mexico has become more prosperous, education levels have increased, and families have become smaller in size. The result is that the large flow of immigration from Mexico to the United States has started to decline. In fact, in 2012, the numbers of Mexicans moving back to Mexico from the United States were greater than those of Mexicans moving to the United States (Passel, Cohn, & Gonzalez-Barrera, 2012; Pew Research Center, 2012).

The status of the various immigrant groups in the United States reflects the global status of their nations of origin and their levels of education. Some Southeast Asian immigrants (such as war refugees from Vietnam, Cambodia, and Laos) came to the United States with little money or education. Today, other vulnerable Asians (such as young women from poor areas) are brought over illegally to work as indentured servants in sweatshops in the Chinatowns of major cities. The vast majority of Asians immigrating to the United States today, however, are educated people with some money who immigrate legally to find greater economic opportunity than exists in their nation of origin. On the other hand, Latin American immigrants (those coming legally, which are the majority) are better able to enter the United States in relatively large numbers without either education or money due to their closer geographical proximity. One result of the disparity in education levels of the various entering immigrant groups is the difference in the positions they achieve in the U.S. workforce and in their subsequent socioeconomic status.

The Social Construction of Race

How do people racially identify you? It depends on where you live as well as what you look like and the racial identity of your ancestors. Race is a social construction, meaning that it is defined differently from society to society and sometimes, over time, even within the same society. For example, the same person could be seen as a member of one racial group in Brazil and another in the United States.

In the United States, some groups have been placed in different racial categories over time. In many ways, Hispanics (meaning Spanish-speakers) and Latino populations (titled Latinx by some, as a gender-neutral term to refer to people of Latin American descent) are treated as a distinct racial group because of their appearance, language, or accent (all of which vary and largely determine the extent to which they face racial discrimination) (Rodriguez, 2000). For example, in the 1930 U.S. Census, Mexican Americans were included under the racial category "Mexican." However, in 1940, they were placed under the racial category "white," "unless they appeared to census interviewers to be 'definitely Indian or of other Nonwhite races' (U.S. Bureau of the Census, 2017b)" (Rodriguez, 2000, p. 84). Today, they are asked to choose a racial category listed on the census and indicate that they are ethnically Hispanic or Latino[3] according to several subcategories. Shifts in U.S. Census racial categories over decades provide a history of how race has been socially constructed differently in different eras.

Hispanics-Latinos/xs are not the first ethnic group to be racialized (where people assign or ascribe constructs of race or ethnicity to others based on appearance) and seen as inferior, however. Many young Italian and Irish Americans may be surprised to learn that most Americans considered members of these ethnic groups to be "less than" white until decades after they arrived in the United States in large numbers.[4] Today, sociologists define a race as a group of people *perceived* to be distinct on the basis of physical appearance, not genetic makeup[5] (Rosenfeld, 2007). Ethnicity refers to cultural rather than physical differences. However, racial and ethnic categories are imperfect and fluid and become complicated for many populations such as Hispanic-Latino/x Americans, who, as noted above, are a distinct ethnic umbrella group but can be of any race.

Today, as Table 8.1 indicates, the racial and ethnic makeup of the United States is still mostly white. However, the percentages of Hispanics-Latinos/xs and Asian Americans are rapidly increasing. Thus, while the term *racial minority* today refers to non-white populations, whites are projected to be a minority racial category by 2045 (Frey, 2018). Table 8.1 indicates the racial demographics of the United States based on 2017 Census figures. A "biracial baby boom" over the three decades since the Supreme Court struck down laws against interracial marriage in 1967 has added to the diversity. By 2015, 10 percent of all marriages and 17 percent of new marriages were interracial (Livingston & Brown, 2017). In 2015, more than 29 percent of Asian and 27 percent of Hispanic American newlyweds married someone of a different race, while 18 percent of black and 11 percent of white newlyweds did so.

Some sociologists who research racial issues maintain that the racial classification system in the United States is changing. George Yancey (2003) and others maintain that Asians and Hispanics-Latinos/xs will eventually "become White." Other scholars, such as Eduardo Bonilla-Silva (2009; Bonilla-Silva & Embrick, 2005), say that the United States is beginning to establish a three-tiered racial hierarchy with "*Whites* ('traditional' whites, new 'white' immigrants; and, in the near future, assimilated Latinos/xs, some multiracials [light-skinned ones], and individual members of other groups [some Asian Americans, etc.])" at the top (Bonilla-Silva & Embrick, 2005, p. 33). Next will come a middle group composed of "*honorary*

Table 8.1	U.S. Racial Demographics (as of 2017)
Race	**Percentage**
White, alone	76.6
White (not Hispanic-Latino)	60.7
Hispanic-Latino	18.1
Black or African American, alone	13.4
Asian, alone	5.8
American Indian or Alaska Native, alone	1.3
Native Hawaiian and Other Pacific Islander, alone	0.2
Two or more races	2.7

Source: U.S. Bureau of the Census (2017)[6]

Note: Based on how people racially identify themselves (total percentages do not equal 100).

Whites" (most light-skinned Latino/x Americans, "Japanese Americans, Korean Americans, Asian Indians, Chinese Americans, the bulk of multi-racials . . . and most Middle Eastern Americans"; Bonilla-Silva & Embrick, 2005, p. 34). The bottom group, the "*collective Black*," will consist of black Americans, dark-skinned Latino/x Americans, Vietnamese Americans, Cambodian Americans, Laotian Americans, "reservation bound" Native Americans, and "maybe Filipino Americans."[7]

No matter what the *future* racial hierarchy will look like, it is important as sociologists to identify that the systems keep reinventing as hierarchies rather than as leveled playing fields and that racial prejudice and inequity continues to significantly exist in the present-day United States. W. E. B. Du Bois (1903/1989) wrote that "the problem of the twentieth century is the problem of the color-line" (p. 35). Unfortunately, the "color-line" (though it may be shifting) is still very much a problem today. In fact, some current scholars maintain that "racism is an integral, permanent, and inde-structible component of this society" (Bell, 1992, p. ix).

In an increasingly diverse society, *race-centered theorists* seek to under-stand how the racial hierarchy adjusts to changing racial and ethnic demo-graphics. Sociologists who use a race-centered theoretical perspective today look at society, as Du Bois did, through the prism of racial inequal-ity. Unlike Marx, they maintain that divisions in society are based on race rather than class.

Race-centered theorists focus on how and why racial hierarchies are actively maintained by those who benefit from them. They point out the glaring gaps in wealth, income, education, incarceration rates, and so forth that benefit whites and disadvantage blacks, Hispanics-Latinos/xs, and Native Americans as some of the many indications of the persistent high levels of racial inequality in the United States. *Critical race theorists*, like Patricia Hill Collins, use a race-centered interdisciplinary perspective to both examine issues of racial inequality and advocate for racial justice. For example, a transgender black woman must deal with gender and racial inequality as well as transphobia, all at the same time. Collins (1990) uses the term *intersectionality* to describe the experience of facing multiple sources of oppression.

Since the 1960s, work by biologist Richard Lewington and others have dispelled the idea that race is genetically determined, allowing for more nuanced understandings of how race is socially constructed. Certainly, though race itself is a social construction, the fact that it exists in our ways of speaking, thinking, and acting imposes very real consequences. Civil rights laws and affirmative action programs aimed to abolish *de jure* (by law) discrimination and provided some opportunities for people of color to rise in socioeconomic status. While these are important, even those policies meant to dismantle racial inequities must be carefully examined, not just based on their de jure intentions but on their *de facto* (in practice) outcomes. For instance, the affirmative action policies passed in the 1960s were designed to eliminate discrimination in employment and education against historically marginalized groups, including women and people of color. However, the group that benefited most has been white women, not women or men of color (Powell, 2008; Crenshaw, 2006; Finley, 2017). Black and Hispanic-Latinx Americans still lag behind most other racial groups in terms of income, wealth, education, and employment. According to the U.S. Bureau of the Census, the median household income in 2017 for Asian Americans was $81,331, and for white, non-Hispanic Americans it was $68,145, but it was just $39,490 for black and $50,486 for Hispanic Americans (Fontenot, K., Semega, J., & Kollar, 2018). Digging into this identifies another disturbing trend, over time, where in 1967 black households earned just 55 percent of white household earnings, in 2016 earnings are still just 61 percent (Saunders, 2017).

These data also illustrate some issues with aggregating data by race, since looking at Asian Americans through this lens we might assume that all are above median income and that the group, as a whole, is doing well. However, a deeper dive identifies that there are four groups of Asian Americans (of the 20 nationalities represented in the United States) with household incomes well below the median household income for all Americans: Bangladeshi ($49,800), Hmong ($48,000), Nepalese ($43,500), and Burmese ($36,000) (López, Ruiz, &

Patten, 2017). Wealth, which requires time to accumulate, reveals even starker disparities between whites and black and Hispanic Americans. The average white American has 10 times the wealth of the average black American and 8.5 times that of the average Hispanic American (Dettling, Hsu, Jacobs, Moore, & Thompson, 2017). While 54.8 percent of Asian Americans and 38.1 percent of non-Hispanic white Americans have completed four years of college, only 23.9 percent of black and 17.2 percent of Hispanic Americans have done so (U.S. Bureau of the Census, 2017a). In May 2018, the unemployment rate was 2.1 percent for Asian Americans, 3.5 percent for white Americans, 4.9 percent for Hispanic-Latino/x Americans, and 5.9 percent for black Americans (Bureau of Labor Statistics, 2019a, Bureau of Labor Statistics, 2019b).

Is this news to you? If you are white, it may well be. The research of many sociologists reveals that "white people are more likely [than black people] to believe that the socioeconomic status of black people is better than it actually is . . . [and that] the playing field is level (Grant-Thomas, as quoted in Valbrun, 2013, para. 16). This ignorance, reflecting a gap between what white Americans *think* is real and what the data show, is partially due to the way race has been depicted in the mass media since the mid-1990s.

The media has a powerful effect on how whites perceive people of color, particularly black Americans, because media outlets are often the only places where most whites see and "get to know" people of color.[8] Relatively few whites (24 percent) live in racially integrated neighborhoods (Ellen, Horn, & O'Regan, 2012). So many tend to formulate their opinions about the socioeconomic status of persons of color (and everything else about other racial groups) from what they see in the media or hear from other secondary sources.

Unfortunately, the media sends a conflicting message. One message is that race-based disparities no longer exist. In magazines, movies, television, and so forth, U.S. society is often portrayed as if race no longer matters. Interracial couples and integrated friendship groups are depicted in advertisements for everything from restaurants to sneakers; popular television shows often have interracial casts that never speak about race (although recent trends with *Grey's Anatomy, This Is Us, Orange Is the New Black*, and *American Crime* show some change); and rarely do news programs or public officials devote time to exposing and analyzing the great racial disparities and systems of institutional racism that continue to persist in the United States today.

Therefore, few white Americans realize that they benefit from *white privilege*. The concept of white privilege refers to the fact that almost every aspect of life (e.g., finding a mate, buying a car, securing a mortgage, shopping for clothes, attaining employment, driving) is measurably easier for white Americans than for Americans of color. This privilege can be "invisible" to white Americans if they have never faced racial discrimination themselves and are disconnected from those who do experience race-based discrimination on a regular basis.[9]

The Color-Blind Ideology

Obama's 2008 speech on race during the Democratic presidential primary will be long remembered because he addressed and contextualized the race-based resentments of both black and white Americans in a straightforward (and eloquent) manner (the speech, *A More Perfect Union*, can be found at https://www.youtube.com/watch?v=zrp-v2tHaDo). His doing so was unprecedented and downright remarkable for a public figure seeking the highest political office in the United States. When Obama won the presidency a few short months later, many commentators declared that his election signaled that we are now in a "postracial" era. However, that is, of course, not the case. His election did not change the fact that racism and racial inequality still exist in the United States. Although the election of an African American president certainly marks a moment of great progress, the points Obama made during his March 2008 speech on race were true in November 2008 and still ring true today. Further, many argue that the election of Donald Trump was a reaction by whites to a perceived threat of non-white Americans "taking over" and the endorsement of a president who visibly represents white privilege.

Many Americans today do not want to hear about the realities of racism and racial inequality. Echoing the media portrayal of race described above, a popular view among many Americans—particularly white Americans—is that race no longer matters. A racial ideology that grew out of conservative arguments against affirmative action programs in the 1980s, the *color-blind ideology* is now the dominant racial ideology in the United States (Bonilla-Silva, 2009; Brunsma, 2006; Powell, 2008; Korgen & Brunsma, 2012). Promoters of the color-blind ideology maintain that we should all act as though we are "color-blind" when it comes to race. However, as President Obama pointed out, this is far from true. Racial tensions, racial inequalities, and racism have existed in the United States for centuries and continue significantly to the present day. Trying to avoid noticing them will not alleviate them.

It is also important to recognize that people's racial experiences influence how they view issues of race. This was painfully apparent after George Zimmerman, a neighborhood watch coordinator in Sanford, Florida, was declared not guilty of shooting and killing Trayvon Martin, an unarmed black teenager walking back to his father's condominium. Patrolling the area for the neighborhood watch organization he led, Zimmerman spotted Martin, made the (racially informed) assumption that he was up to no good, followed him, got out of his car to pursue him on foot, became involved in a fistfight with Martin, and then pulled out his gun and shot him during the fight.

The vast majority of black Americans were outraged when Zimmerman was found not guilty of second-degree murder and manslaughter, but many white Americans were not. According to a Pew Research Center (2013) poll taken shortly after the trial, 86 percent of black but just 30 percent of white respondents were dissatisfied with the verdict (21 percent of whites and 9 percent of blacks declared that they "didn't know" if they were satisfied or dissatisfied).

Research shows that people notice the racial appearance of those around them almost instantly (Apfelbaum, 2012). However, the influence of the color-blind ideology has led most people to feel uncomfortable talking about race, particularly with members of another race (those most likely to provide us with a different perspective). Even blacks and whites who are close friends tend to avoid talking about racial issues (Korgen, 2002). While all Americans are affected (at least somewhat) by the color-blind ideology (Bonilla-Silva, 2009), white Americans, who do not tend to be confronted by racism in their day-to-day lives, are particularly susceptible to its influence. White Americans also tend to be particularly uncomfortable talking about racism, a phenomenon Robin DiAngelo has termed "white fragility," which can prevent productive dialogue and action toward racial justice (DiAngelo, 2011).

The negative repercussions of acting as if we do not notice race are many. For example, trying to ignore race in the workplace can result in less productive management styles and a lack of trust among coworkers (Apfelbaum, 2012). Not acknowledging the race of your friends can prevent you from understanding how race and racism affect their lives, their perceptions of themselves, and the society in which you all live. This avoidance is problematic for anyone interested in promoting racial justice. If we want to end racial discrimination, we first have to acknowledge that it exists.

Effective efforts to fight racial discrimination can come only after we uncover and confront patterns of racial discrimination. Likewise, we must make the "invisible privileges" (Rothenberg, 2004) of whiteness visible. To do so, we need to notice and keep track of how different racial groups are treated. For example, if we did not do so, we would not know that in Milwaukee, Wisconsin, and New York City, among equally qualified white and black job applicants, whites were more than twice as likely to be offered a second interview. Even criminal records did not erase the advantage white applicants had over black applicants. Whites with criminal records were just as likely as, if not more likely than, black applicants *without* a criminal record to get a second interview (Pager, 2008).

Nor would social scientists (and, in turn, the general public) be able to learn that New York City had a "pattern, practice and policy of intentional discrimination against black applicants" to the fire department (Baker, 2010, para. 2). Nor would we know that black and Hispanic emergency room patients are less likely to receive medicine to reduce pain than white patients (Heins et al., 2006; O'Leary, 2013; Terrell et al., 2010) or that Asian, Hispanic, and black customers were charged more than whites for auto insurance (Consumer Financial Protection Bureau, 2013).

If we did not keep track of racial patterns, we, likewise, would be unaware of environmental racism. Robert D. Bullard, known as the "Father of Environmental Justice," has conducted sociological research since the late 1970s to uncover and lead campaigns against environmental racism. His work has helped bring to public attention the fact that people of color suffer disproportionately from environmental hazards and that governmental responses to emergencies differ based on the race of the citizens affected (Bullard, 2008). Bullard's award-winning book, *Dumping in Dixie* (Bullard, 2000), remains essential reading for all those interested in learning about environmental racism and working for environmental justice. A member of the National Environmental Justice Advisory Council, Bullard participated in drafting President Clinton's Executive Order 12898, which ordered every federal agency to make environmental justice a part of its mission and has been an indefatigable advocate for environmental justice. (It's important to note that under the Trump administration, there is widespread concern that the Environmental Justice Advisory Council is being altered to move away from the science and research that it is intended to be guided by). Bullard's work and the studies mentioned above (and the myriad more that exist) illustrate conclusively that race *does* matter and that we cannot be color-blind when it comes to creating and implementing policy in the United States.

The color-blind perspective on race also runs counter to the discipline of sociology. It undermines two of the most powerful goals of sociology: (a) to observe how society really works and (b) to give voice to the marginalized groups within it. It is our obligation as sociologists to expose the negative repercussions of the color-blind ideology and to support efforts to promote racial justice.

Racism Is a Global Issue

Racism is not just a U.S. problem. The concepts of race and ethnicity have been used to separate and distinguish groups of people from all over the globe and to justify histories of colonialism and conquest. Wherever you go in the world, you will find, to varying degrees, some racial and ethnic tension. Western Europe provides a good example of past and present racial tensions. Historically, the people of western Europe have suffered many wars among their nations that were, at least in part, due to racism and *ethnocentrism* (the belief that one's ethnic or racial group is superior to all others). In fact, the leaders of Nazi Germany deemed the Aryan race to be the "ideal race" and Jewish people to make up a race that must be eradicated. They succeeded in killing six million Jews in history's largest genocide and made western Europe a place where few Jewish people remain.

Over the past quarter century, western Europe has become more racially and ethnically diverse due to the arrival of increasing numbers of Asian, Middle Eastern, and African immigrants searching for work. With this diversity have come new types of racial and ethnic tension. When plenty of jobs exist and immigrants are needed to take the ones that the citizens of those nations do not want, immigrants tend to be more welcome. However, when jobs become scarce and citizens feel that they must compete with others for them, immigrants can become easy targets. The different races and religions of many of the immigrants in European nations have helped spark anger toward and sometimes violence against non-white and non-Christian immigrants.

Evidence of racism and ethnocentrism also abounds in soccer stadiums across Europe. Today, "almost every country in Europe has racist signs, chants and even violence at soccer stadiums, particularly from rightist groups that single out blacks, Jews, Muslims or other ethnic groups" (Vecsey, 2003, Section 8, p. 1). Indeed, racism has been described as the "scourge" of soccer stadiums in Europe (Bohlen, 2013; Pugmire, 2009). A 2012 BBC documentary included footage of fans in Ukraine and Poland "giving the Nazi salute, making monkey chants, and displaying anti-Semitic behavior" and other fans viciously attacking students from Asia attending a match in a Ukrainian soccer stadium (Longman, 2012). The following Sociologist in Action section describes one college student's encounter with racism in Italy.

Sociologist in Action

Michael Cermak

Michael Cermak uses Participatory Action Research (PAR) and creative action media projects to partner with youth and challenge racial biases underlying the food and environmental movements. When he was a PhD student studying the sociology of urban environmental education, Cermak began a collaborative research project

(Continued)

(Continued)

with black and Latinx youth who had struggled at the margins of educational institutions, some of whom had dropped out. As Cermak describes, "As a scholar of color who cares deeply about urban sustainability and youth empowerment, I wanted to do more than just describe the racial bias in yet another progressive movement and intervene by working with a set of teens to create food media that represented more of their perspective."

Over six months, they ran workshops focused on food justice issues and considered the other social justice issues they faced, like community violence and structural racism. In the ensuing six months, the group studied the art of filmmaking and collaboratively directed and produced *Planting for Peace: Bury Seeds, Not Bodies* (the film can be found at https://vimeo.com/31636790). The film challenges urban food landscapes, like food deserts that have a total lack of fresh and whole foods, and advocates for green jobs for urban youth.

Once the film was complete, Cermak and his youth partners toured New England, screening the film and providing workshops over 75 times and using the film as a platform for advocacy. Through this process, some of the youth gained employment as public speakers, and the collective created two raised-bed gardens that shifted concepts of youth participation and community ownership in urban spaces. The project showed the potential of using creative media to advance change. As Cermak shares, "I would never have thought of myself as a filmmaker, but becoming a sociologist in action required that I use any medium at my disposal to amplify the voices that needed to be heard."

Note: Quotes excerpted from Cermak, Michael. 2015. "Amplifying the Youth Voice of the Food Justice Movement with Film: Action Media Projects and Participatory Media Production," in *Sociologists in Action on Inequalities: Race, Class, Gender and Sexuality.* SAGE.

Exercise 8.1
Immigration, Opportunity, and Fair Trade

1. Read "Mexican Coffee Cooperative With US Customers Helps Build Bridges" (https://tucson.com/news/local/mexican-coffee-cooperative-with-u-s-customers-helps-build-bridges/article_280a6a35-46f2-5a1d-a838-cd57f542aff7.html).

2. Read through the resources on the website for Café Justo at http://www.justcoffee.org/ and make sure to read the Fact Sheet (http://www.justcoffee.org/wp-content/uploads/2014/06/factsheet.pdf). Then, watch the documentary "Just Coffee: Small Solutions for Global Problems" (https://www.youtube.com/watch?v=Y8anKFz53fM&feature=player_embedded).

Then, write a two-page essay that describes the work of Café Justo, how it relates to immigration and opportunity, what it teaches you about fair trade and direct trade, and how it influences your thinking on border policy and politics. Note: if you want to learn even more, you can order the book *Just Coffee: Caffeine With a Conscience* (http://www.justcoffee.org/product/just-coffee-the-book/).

Exercise 8.2
Africa and the Legacy of Colonialism

At the same time the civil rights movement was under way in the United States, a global movement for black rights was taking place. One of the results of this global movement was the end of colonial rule in Africa. (All of Africa, except Liberia and Ethiopia, had been colonized by European nations.)

1. Go to the Library of Congress's Country Studies website (http://lcweb2.loc.gov/frd/cs).

2. Select an African nation (other than Liberia or Ethiopia), and write a background paper that includes information about (a) when it was colonized, (b) what nation colonized it, (c) what conditions were like under the rule of the colonizers, (d) when it became a free nation, (e) how its borders were established, (f) how the establishment of its borders affects the nation today, and (g) the overall legacy of colonialism in the nation today.

Extra credit: Analyze the current social, political, and economic well-being of the country you have chosen. How has globalization helped or hindered the nation's growth and the overall quality of life of the citizens? Look at the website https://www.hrw.org/countries or https://www.amnesty.org/en/countries/africa/ for more information and insight.

Exercise 8.3
Myths and Facts About Immigration

1. Read the article "Myths and Facts About Immigrants and Immigration" at https://www.adl.org/resources/fact-sheets/myths-and-facts-about-immigrants-and-immigration and "Immigration Myths and Facts" from the U.S. Chamber of Commerce (https://www.uschamber.com/sites/default/files/documents/files/022851_mythsfacts_2016_report_final .pdf).

2. What did you learn? Did you hold any of the myths? Do the facts change your mind? Why do you think Americans hold such deep but incorrect myths about immigration? What are the consequences of this?

3. Create a one- or two-page "fact sheet" that you can share with your campus, hometown community, family, and friends. On the fact sheet, be sure to also include at least three resources for them to do further research and three actions they can take in order to work against the prevailing myths in our society on the issues of immigrants and immigration.

Exercise 8.4
Survey on Race in the United States

Conduct a survey of the members of one of your classes (not the course for which you're reading this book). On the survey, ask respondents to respond to the following statements and questions:

(Provide them with the following options for Statements 1 to 3 and ask them to circle one to indicate their level of agreement: Strongly Agree, Somewhat Agree, Not Sure, Somewhat Disagree, Strongly Disagree.)

1. The socioeconomic status of whites and people from other races in the United States is relatively equal.

2. Racism is often exaggerated by people of color.

3. People of all races have relatively equal chances to become successful in the United States today.

4. What, if any, college courses have you taken that deal with race and ethnicity, racism, and racial inequalities?

5. What are your race and your ethnicity?

Compare the answers of (a) respondents from different racial and ethnic groups and (b) those who have and those who have not taken any courses that deal with racial and ethnic relations. Analyze your results.

Exercise 8.5
Racism in Sports

Watch "Beautiful Game Turned Ugly: Racism in Europe's Soccer Arenas" at http://www.youtube.com/watch?v=W-iRLmaZf4A and watch "Sol Campbell Warns Fans to Stay Away From Euro 2012" (http://www.bbc.co.uk/news/uk-18240143).

Based on the information in the videos and what you have read in this chapter, answer the following questions:

1. Were you surprised by the information in the videos? Why or why not?

2. If you were the commissioner of FIFA (the international football [soccer] federation), what would you do to curb the racist behavior at the stadiums? What makes you think your efforts would succeed (or not)?

3. How might addressing racial inequality in European nations be more or less difficult than addressing it in the United States?

Exercise 8.6
Has Martin Luther King Jr.'s Dream Been Realized?

1. Go to the website http://www.holidays.net/mlk/speech.htm

2. Read and watch Dr. Martin Luther King Jr.'s "I Have a Dream" speech.

3. List the points King makes about when he will be "satisfied" with the situation for black people in America.

4. Check off those goals that have now been achieved. Provide evidence for your decisions to check off or not check off each point. Some of these may be partially achieved, whereas others may have been fully achieved or not at all.

5. Read the one-pager from the Campaign to End Racial Profiling at http://www.nea.org/assets/docs/RP_One_Pager_6-12-12_Draft.pdf and "On Views of Race and Inequality, Blacks and Whites are Worlds Apart" at http://www.pewsocialtrends.org/2016/06/27/on-views-of-race-and-inequality-blacks-and-whites-are-worlds-apart/ and listen to "The Legacy of the 'Little Rock Nine'" (http://www.npr.org/templates/story/story.php?storyId=14692397).

6. Based on all of the above, do you think Martin Luther King Jr. would be satisfied with the status of people of color today? Why or why not? If not, how do you suggest that we move toward a fuller realization of his dream? What specific policy recommendations would you make?

Exercise 8.7
White Privilege

Peggy McIntosh did much to bring white privilege to public recognition with her essay "White Privilege: Unpacking the Invisible Knapsack" (1989), in which she listed some of the "daily effects of White privilege" on her life.

If you are white, come up with a list of approximately 10 privileges that you (personally) enjoy for being white. (Do not, for example, state that you would have an easier time getting a mortgage unless you actually have a mortgage and you had an easy time getting it.) If you are a person of color, come up with a list of approximately 10 privileges that you think a white person receives for being white that you do not have.

Compare your answers with those of other members of the class.

1. What are the most significant differences among the lists?

2. Who do you think had the most difficult time coming up with the 10 privileges? Who had the easiest time? Why?

Extra Credit: Read "White Fragility" by Robin DiAngelo (https://libjournal.uncg.edu/ijcp/article/viewFile/249/116). What is white fragility? How do you encounter white fragility in your own life? If you are white, how might you experience this personally? Are there times when you yourself have been guided by white fragility? What does DiAngelo suggest for overcoming white fragility, and what can you do to help yourself and others toward that goal?

Exercise 8.8
White Privilege: Creating Advantages and Disadvantages

1. Fill out the "White Privilege Checklist" found at http://crc-global.org/wp-content/uploads/2012/06/white-privilege.pdf (Note: fill this form out no matter which race you identify with).

2. Read the following four short articles: "Study: Anti-Black Hiring Discrimination Is as Prevalent Today as It Was in 1989" at https://www.vox.com/identities/2017/9/18/16307782/study-racism-jobs, "Black Men Get Longer Prison Sentences Than White Men for the Same Crime: Study" at https://abcnews.go.com/Politics/black-men-sentenced-time-white-men-crime-study/story?id=51203491, "Blacks and Latinos Denied Mortgages at Rates Double Whites" at http://www.wunc.org/post/blacks-and-latinos-denied-mortgages-rates-double-whites, and "The Whiter the Neighborhood, the Greater the Penalty Black Renters Pay, Study Shows" (https://www.cbsnews.com/news/the-whiter-the-neighborhood-the-greater-the-penalty-black-renters-pay/).

3. Write a two- to three-page paper outlining and analyzing what you have read and attempting to contextualize it inside the concept of white privilege, including results from your own White Privilege Checklist.

Exercise 8.9
Your Family and Issues of Race

Write a one- to two-page paper that answers the following questions:

1. When you were growing up, what did you hear about race and racism from your family? (For example, were race and/or racism ever seriously discussed? If so, why and how often was race a topic of serious discussion? Did your family tend to bring up race only when making jokes or insulting comments about members of other races? Did they speak from a viewpoint of color-blindness and thus discourage any discussion of the real, pressing issues associated with race and racism?)

2. How do you think the racial makeup of your family influenced how race was discussed (or not discussed) in your family?

3. How do your answers to the first two questions relate to what you have learned in this chapter? How does the information in this chapter help you, if at all, to reevaluate your views on race, immigration, and racism?

Exercise 8.10
"A More Perfect Union"

Go to http://www.youtube.com/watch?v=pWe7wTVbLUU and watch Barack Obama deliver his "A More Perfect Union" speech.

Write a two- to three-page paper that describes (a) the major points Obama makes in the speech, (b) how they relate to what you learned in this chapter (making sure to address how this speech challenges the color-blind ideology), and (c) your reaction to the speech. (If you heard the speech when it was first given on March 18, 2008, compare your reaction to it then and your reaction to it now, after reading this chapter.)

Exercise 8.11
De Facto Segregation: Not So Neighborly

1. Read "America Is Diverse as Ever, but Still Segregated" at https://www.washingtonpost.com/graphics/2018/national/segregation-us-cities/?utm_term=.fb021ea73aa5 and "Black-White Segregation Edges Downward Since 2000" (https://www.brookings.edu/blog/the-avenue/2018/12/17/black-white-segregation-edges-downward-since-2000-census-shows/).

2. Look at and play around with the map on National Geographic's "Where We Live, Block by Block" (https://www.nationalgeographic.com/magazine/2018/10/diversity-race-ethnicity-united-states-america-interactive-map/).

3. Write a two-page paper analyzing *de facto* housing segregation in the United States today.

Exercise 8.12
Environmental Racism

Environmental racism refers to a type of environmental inequality that negatively affects people of color. An examination of environmental inequality in the United States reveals evidence of environmental racism. For example, Americans of color are more likely than white Americans to be exposed to a variety of pollutants and poisons in their homes, neighborhoods, and workplaces.

Read "Black Lives Matter: Environmental Racism Is Killing African-Americans" at https://blackamericaweb.com/2017/11/16/black-lives-matter-environmental-racism-is-killing-african-americans/, and "Silent Discrimination: Issues of Environmental Justice" at http://www.mindthesciencegap.org/2012/01/16/silent-discrimination-issues-of-environmental-justice/, and "What Standing Rock Teaches Us About Environmental Racism and Justice" at

(Continued)

(Continued)

https://www.healthaffairs.org/do/10.1377/hblog20170417.059659/full/, and then answer the following questions in a two- to three-page paper:

1. What is environmental racism? Had you heard much about it before? If not, why not? If so, what had you heard, and what have you learned from these readings?

2. Why are Americans of color more likely than white Americans to be exposed to pollutants?

3. How do you think your own racial and class background influence your reaction to the information in these articles?

4. How might you devise a plan to use the information in these articles and the rest of the chapter to mobilize students on your campus to organize against environmental racism in your area?

DISCUSSION QUESTIONS

1. Close your eyes and picture a U.S. citizen. What does the person you pictured look like? Why do you think you imagined the race and ethnicity of the person in the way you did? (It might be interesting to ask people who are not in this course the same question after you have answered for yourself. If you ask them only to "describe" the person, note whether they mention the race.)

2. Do you have a good friend who is of a different race from yourself? If so, how did you meet? Do you ever talk seriously about racial issues? Why or why not? Do you think you understand the impact of race on your friend(s)? Why or why not? If you do not have a good friend of another race, why do you think that is?

3. How can sociology be used to (a) recognize, (b) publicize, and (c) combat racial discrimination and racism? Be specific and clearly explain your answers.

4. Why do you think so many people want to believe that "race doesn't matter anymore"? What do you think would happen if we no longer kept track of different racial groups in our society? What might be the beneficial effects? What might be the negative consequences?

5. What do you think your parents would say if you told them you were going to marry someone of a different race? Why? Would it depend on what race? Why?

6. Which groups of prospective students (other than racial groups) do you think have an easier time gaining admittance to and paying for a college education than other groups of prospective students? Why do you think people have a harder time accepting race-based affirmative action programs than the preferential treatment given to other, non-race-based groups?

7. Do you think multiracial Americans should be given a separate "multiracial" box to check on the U.S. Census? What do you think would be some outcomes of the establishment of such a category?

8. What do you think Marx would say is the root cause of racial discrimination? Do you agree? Why or why not?

SUGGESTIONS FOR SPECIFIC ACTIONS

1. Conduct interviews with administrators at your school to find out (a) what they think is the obligation of an institution of higher education regarding combating racism in society and (b) what specific things your school is doing to combat racism. If you think the school should be doing more, organize a group of like-minded students, faculty, and staff to create more antiracism efforts on campus. Administrators can range from president of the university to director of the campus multicultural affairs office, and so forth.

2. Go to the NAACP website at www.naacp.org. Look over the volunteer opportunities listed there and find one you are interested in pursuing. Find the number of your local NAACP chapter (by calling 877-NAACP-98) and offer your services.

   Or join or volunteer with another race and/or ethnicity-based organization (e.g., MANA at http://www.hermana.org, League of United Latin American Citizens at http://lulac.org, CAPAL [Conference on Asian Pacific American Leadership] at https://www.capal.org/site/, or the American Indian Movement at http://www.aimovement.org).

   Please go to this book's website at http://study.sagepub.com/white6e to find further civic engagement opportunities, resources, peer-reviewed articles, and updated web links related to this chapter.

NOTES

1. Throughout the book, we use the terms Latino and Latinx together or interchangeably, as these terms are under debate as of the writing of this book. The debate revolves around the gendered nature of the Spanish language using the masculine "Latino" to describe both men and all people, and the feminine "Latina" to describe women only.
2. The quotas were based on the U.S. population back in 1890, when the numbers of darker-skinned Europeans (e.g., Italians) were lower than they were in 1924.
3. Although we have noted that ethnicity and race are two distinct concepts, we group them together in several places in this chapter. We typically do so when we are including Hispanics-Latinos/xs in the discussion, because they are an *ethnic* group that tends to be viewed—by its own members, the U.S. Census, and by other Americans—as a distinct *racial* group.
4. In the middle decades of the 1800s, Irish Americans were commonly portrayed in newspaper and magazine cartoons as apes.
5. Thanks to the Human Genome Project, we now know that physical differences vary more within than among races (Jorde & Wooding, 2004).
6. This chart uses Census language and categories, which are not reflective of current debates on categories, including that of Latino/x.
7. Bonilla-Silva and Embrick (2005) acknowledge that some individual members of these groups may fall outside the designated tier.
8. Black Americans are more isolated than either Hispanic or Asian Americans (see Charles, 2003).
9. For a great discussion of "invisible privileges," see Paula Rothenberg's (2004) *Invisible Privilege: A Memoir About Race, Class, and Gender.*

REFERENCES

Adelman, L. (Producer), Herbes-Sommers, C., Strain, T. H., & Smith, L. (Directors). (2003). *Race: The power of an illusion* [Documentary series]. USA: California Newsreel.

Apfelbaum, E. P. (2012, January 31). *Opinion: The risks of ignoring race in the workplace.* Retrieved from CNN website at http://edition.cnn.com/2012/01/31/opinion/apfelbaum-colorblind/

Baker, A. (2010, January 13). Judge cites discrimination in N.Y. fire dept. *The New York Times.* Retrieved from http://www.nytimes.com/2010/01/14/nyregion/14fire.html

Bell, D. (1992). *Faces at the bottom of the well.* New York, NY: Basic Books.

Bohlen, C. (2013, January 3). A struggle with identity and racism. *The New York Times.* Retrieved from http://www.nytimes.com/2013/01/05/world/europe/05iht-letter05.html?_r=0

Bonilla-Silva, E. (2009). *Racism without racists: Color-blind racism and the persistence of racial inequality in America* (3rd ed.). Lanham, MD: Rowman & Littlefield.

Bonilla-Silva, E., & Embrick, D. G. (2005). Black, honorary white, white: The future of race in the United States. In D. Brunsma (Ed.), *Mixed messages: Multiracial identities in the color-blind era* (pp. 33–48). Boulder, CO: Lynne Rienner.

Brunsma, D. (2006). *Mixed messages: Doing race in the color-blind era.* Boulder, CO: Lynne Rienner.

Bullard, R. D. (2000). *Dumping in Dixie: Race, class and environmental quality* (3rd ed.). Boulder, CO: Westview Press.

Bullard, R. D. (2008). Differential vulnerabilities: Environmental and economic inequality and government response to unnatural disasters. *Social Research, 75*(3), 753–784.

Bureau of Labor Statistics. (2019a). Table A-2. Employment status of the civilian population by race, sex, and age. Retrieved from https://www.bls.gov/news.release /empsit.t02.htm

Bureau of Labor Statistics. (2019b). Table A-3. Employment status of the Hispanic or Latino population by sex and age. Retrieved from https://www.bls.gov/news. release/empsit.t03.htm

Cermak, M. (2015). Amplifying the youth voice of the food justice movement with film: Action media projects and participatory media production. In *Sociologists in action on inequalities: Race, class, gender and sexuality.* Thousand Oaks, CA: Sage.

Charles, C. (2003). The dynamics of racial residential segregation. *Annual Review of Sociology, 29,* 167–207

Collins, P. H. (1990). *Black feminist thought: Knowledge, consciousness, and the politics of empowerment.* New York, NY: Routledge.

Consumer Financial Protection Bureau. (2013, December 20). *CFPB and DOJ order Ally to pay $80 million to consumers harmed by discriminatory auto loan pricing.* Retrieved from http://www.consumerfinance.gov/newsroom/cfpb-and-doj-order-ally-to-pay-80-million-to-consumers-harmed-by-discriminatory-auto-loan-pricing/

Crenshaw, K. W. (2006). Framing affirmative action. *Michigan Law Review First Impressions, 105*(1), 123–133.

DiAngelo, R. (2011). White fragility. *The International Journal of Critical Pedagogy, 3*(3).

DeNavas-Walt, C., Proctor, B. D., & Smith, J. C. (2013, September). *Income, poverty, and health insurance coverage in the United States: 2012.* Retrieved from https:// www.census.gov/prod/2013pubs/p60-245.pdf

Dettling, L. J., Hsu, J. W., Jacobs, L., Moore, K. B., & Thompson, J. P. (2017). *Recent trends in wealth-holding by race and ethnicity: Evidence from the Survey of Consumer Finances.* Retrieved from the Federal Reserve website at https://www.federalreserve .gov/econres/notes/feds-notes/recent-trends-in-wealth-holding-by-race-and-ethnicity-evidence-from-the-survey-of-consumer-finances-20170927.htm

Du Bois, W. E. B. (1989). *The souls of black folk.* New York, NY: Penguin Books. (Original work published 1903)

Ellen, I. G., Horn, K., & O'Regan, K. (2012, May). *Pathways to integration: Examining changes in the prevalence of racially integrated neighborhoods*. Retrieved from http://furmancenter.org/files/publications/Pathways_to_Integration_May_2012_2.pdf

Finley, T. (2017, August 3). 4 key points that debunk misconceptions around affirmative action. Retrieved from https://www.huffingtonpost.com/entry/affirmative-action-still-matters_us_5981d9b6e4b0353fbb33e1bb

Fontenot, K., Semega, J., & Kollar, M. (2018, September 12). *Income and poverty in the United States: 2017*. Retrieved from United States Census Bureau website at https://www.census.gov/library/publications/2018/demo/p60-263.html

Frey, W. H. (2018, March 14). The US will become 'minority white' in 2045, Census projects. Retrieved from https://www.brookings.edu/blog/the-avenue/2018/03/14/the-us-will-become-minority-white-in-2045-census-projects/

Gibson, C. J., & Lennon, E. (1999). *Historical census statistics on the foreign-born population of the United States: 1850 to 1990* (Working Paper No. 29; U.S. Bureau of the Census, Population Division). Retrieved from http://www.census.gov/population/www/documentation/twps0029/twps0029.html

Heins, J. K., Heins, A., Grammas, M., Costello, M., Huang, K., & Mishra, S. (2006, June). Disparities in analgesia and opioid prescribing practices for patients with musculoskeletal pain in the emergency department. *Journal of Emergency Nursing*, 32(3), 219–224.

Henderson, T. (2018, September 20). Drop in Mexican-born immigrants attributed to hostility here, opportunity there. Retrieved from https://www.pewtrusts.org/en/research-and-analysis/blogs/stateline/2018/09/20/drop-in-mexican-born-immigrants-attributed-to-hostility-here-opportunity-there

Jorde, L. B., & Wooding, S. P. (2004). Genetic variation, classification, and "race." *Nature Genetics*, 36, S28–S33. Retrieved from http://www.nature.com/ng/journal/v36/n11s/full/ng1435.html

Korgen, K. O. (2002). *Crossing the racial divide: Close friendships between black and white Americans*. Westport, CT: Praeger.

Korgen, K. O., & Brunsma, D. (2012). Avoiding race or following the racial scripts? Obama and race in the recessionary part of the colorblind era. In A. Jolivette (Ed.), *Obama and the biracial factor: The battle for a new American majority* (pp. 191–204). Bristol, England: Policy Press.

Lange, J., & Torbati, Y. (2018, September 13). U.S. foreign-born population swells to highest in over a century. Retrieved from https://www.reuters.com/article/us-usa-immigration-data/u-s-foreign-born-population-swells-to-highest-in-over-a-century-idUSKCN1LT2HZ

Lazarus, E. (2009). *The new colossus* (Academy of American Poets). Retrieved from http://www.poets.org/viewmedia.php/prmMID/16111. (Original work published 1883)

Livingston, G., & Brown, A. (2017). Intermarriage in the US 50 years after Loving v. Virginia. Retrieved from Pew Research Center website at http://www.pewsocial-trends.org/2017/05/18/intermarriage-in-the-u-s-50-years-after-loving-v-virginia/

Longman, J. (2012, May 30). Racism and soccer are in play at a big event in east Europe. *The New York Times*. Retrieved from http://www.nytimes.com/2012/05/31/sports/soccer/concerns-of-racism-precede-european-soccer-championships.html?page wanted=all

López, G., Ruiz, N. G., & Patten, E. (2017, September 8). Key facts about Asian Americans, a diverse and growing population. Retrieved from http://www.pew research.org/fact-tank/2017/09/08/key-facts-about-asian-americans/

McIntosh, P. (1989, July). *White privilege: Unpacking the invisible knapsack*. Retrieved from http://www.isr.umich.edu/home/diversity/resources/white-privilege.pdf

O'Leary, L. (2013, September 13). How race can impact your health care. *Marketplace Money*. Retrieved from http://www.marketplace.org/topics/your-money/health-care/how-race-can-impact-your-health-care

Pager, D. (2008, August 9). *Study: Black man and white felon—same chances for hire*. Retrieved from CNN website at http://ac360.blogs.cnn.com/2008/08/09/study-black-man-and-white-felon-same-chances-for-hire

Passel, J., Cohn, D., & Gonzalez-Barrera, A. (2012, June 1). *Net migration from Mexico falls to zero—and perhaps less*. Retrieved from Pew Hispanic Center website at http://www.pewhispanic.org/2012/04/23/net-migration-from-mexico-falls-to-zero-and-perhaps-less

Pew Research Center. (2012, June 19). *The rise of Asian Americans*. Retrieved from http://www.pewsocialtrends.org/files/2012/06/SDT-The-Rise-of-Asian-Americans-Full-Report.pdf

Pew Research Center. (2013, July 22). *Big racial divide over the Zimmerman verdict*. Retrieved from http://www.people-press.org/files/legacy-pdf/7-22-13%20Zimmerman%20Trial%20Release.pdf

Pierce, S., & Selee, A. (2017). Immigration under Trump: A review of policy shifts in the year since the election (policy brief). Retrieved from Migration Policy Institute website at https://www.migrationpolicy.org/sites/default/files/publications/TrumpatOne-final.pdf

Powell, J. A. (2008). Post-racialism or targeted universalism. *Denver University Law Review, 86*(3), 785–806.

Pugmire, J. (2009, February 2). Little done to stop racism in European soccer. *USA Today*. Retrieved http://www.usatoday.com/sports/soccer/2009-02-07-1089560368_x.htm

Rodriguez, C. E. (2000). *Changing race: Latinos, the census, and the history of ethnicity in the United States*. New York, NY: New York University Press.

Rosenfeld, M. J. (2007). *Age of independence: Interracial unions, same-sex unions, and the changing American family*. Cambridge, MA: Harvard University Press.

Rothenberg, P. (2004). *Invisible privilege: A memoir about race, class, and gender*. Lawrence: University Press of Kansas.

Saunders, P. (2017, December 8). Where African-American incomes are rising—and where they're not. *Forbes*. Retrieved from https://www.forbes.com/sites/petesaunders1/2017/12/08/where-african-american-incomes-are-rising-and-where-theyre-not

Terrell, K. M., Hui, S. L., Castelluccio, P., Kroenke, K., McGrath, R. B., & Miller, D. K. (2010, July). Analgesic prescribing for patients who are discharged from an emergency department. *Pain Medicine, 11*(7), 1072–1077.

United States Elections Project. (2018, December 14). *2018 November general election turnout rates*. Retrieved from http://www.electproject.org/2018g

U.S. Bureau of the Census. (2016). *QuickFacts*. Retrieved from https://www.census.gov/quickfacts/fact/table/US/PST045217

U.S. Bureau of the Census. (2017a, September 12). Income, poverty and health insurance coverage in the United States: 2016. Retrieved from https://www.census.gov/newsroom/press-releases/2017/income-povery.html

U.S. Bureau of the Census. (2017b). CPS historical time series tables [Data file and code book]. Retrieved from https://www.census.gov/data/tables/time-series/demo/educational-attainment/cps-historical-time-series.html

Valbrun, M. (2013, August 19). Widespread bias continues in America despite claims of post-racial society. *America's Wire* (Maynard Center on Structural Inequity). Retrieved from http://americaswire.org/drupal7/?q=content/widespread-bias-continues-america-despite-claims-post-racial-society

Vecsey, G. (2003, February 2). SOCCER: England battles the racism infesting soccer. *The New York Times*, Section 8, p. 1. Retrieved from http://www.nytimes.com/2003/02/02/sports/soccer-england-battles-the-racism-infesting-soccer.html?pagewanted=all&src=pm

Wang, W. (2012, February 16). *The rise of intermarriage rates: Rates, characteristics vary by race and gender*. Retrieved from Pew Research Center website at http://www.pewsocialtrends.org/2012/02/16/the-rise-of-intermarriage/2/#chapter-1-overview

Wellman, D. T. (1993). *Portraits of white racism*. Cambridge, England: Cambridge University Press.

Yancey, G. (2003). *Who is white? Latinos, Asians, and the new black/nonblack divide*. Boulder, CO: Lynne Rienner.

9

Sex, Gender, and Power

I magine you have just found out that you are going to be a parent. What are your hopes and dreams for your child? What games will you play with your little one? Can you picture yourself as the coach of one of your child's sports teams? Can you imagine you and your child baking cookies together? Now, think about some of your child's characteristics. Is your child going to be tough? Sensitive? A leader? A follower? More than likely, these questions are difficult for you to answer without first picturing whether your child will be a boy or a girl.

Although the physical characteristics and genetic makeup of girls and boys play a major role in the adults they will become, so do the gender roles assigned to them. All parents see their children through glasses tinted by gender socialization. Through gender socialization, we learn to consciously and subconsciously apply dissimilar social roles to boys and girls. The result is that we treat boys and girls differently, and they, in turn, learn to act in "masculine" or "feminine" ways. In this chapter, we examine how gender is socially constructed, how gender construction relates to the distribution of power in society, the status of women in the United States, and how we might address the inequality of power between men and women.

The Social Construction of Gender

A person's sex is determined by the physical characteristics that distinguish males and females,[1] whereas gender is determined by the social roles assigned to males and females in society. As such, gender is socially constructed. Sex differences generally remain constant,[2] while gender differs over time and from society to society. For example, in the mid-1900s, there were more male than female doctors in the United States, but women dominated the medical field in the Soviet Union. Women in Saudi Arabia only recently won the legal right to vote and drive a vehicle and are still subject to guardianship by a male family member, whereas in much of the rest of the world, rights to vote and drive and to self-determination have been legal for women for a long time (Human Rights Watch, 2016; Sirgany & Smith-Spark, 2018; Coker, 2018).

Here in the United States, until recently, it was seen as inappropriate for girls to play ice hockey because it was thought to be too physical a game for them and was considered unfeminine. However, over the past 25 years, as women's roles and the idea of femininity have been reconstructed, the number of girls and women who have registered with USA Hockey has jumped from 10,000 in 1992 through 1993 to 79,355 in 2017 through 2018 (USA Hockey, 2019).

As the discussion above illustrates, gender has also been traditionally seen as *binary*—that is, having only two distinct possibilities of male or female, which are assumed to align with a person's sex. However, as you may know from personal experience, friendships, or by following Caitlin Jenner's transition to female from her prior identity as Bruce Jenner, gender may change. A person who assumes the gender identity associated with their sex at birth is *cis-gender*, whereas a *transgender* person's identity does not align with their sex at birth, and they may undertake a gender transition that involves a change of name, self-presentation, and sometimes hormone therapy and/or surgery (Zevallos, 2019).[3] Have you ever introduced yourself or met others in social settings and shared personal pronouns? Understanding that gender is socially constructed and that people may be in transition, this practice invites people to self-identify rather than allow assumptions about gender to be based on appearance. People may self-identify along binary terms, using *she, her, hers* or *he, him, his* pronouns or use *third gender* or non-binary, pronouns, which may include *they, them, theirs* or other options.

Although people's personalities, talents, and outlooks on life are based on a combination of genetic (nature) and social (nurture) factors, sociologists focus on the social factors. In the case of gender and sexuality, we study the social roles applied to men and women through gender socialization. While doing so, it is important to remember that gender socialization intersects with how people deal with race, ethnicity, age, sexual orientation, and so forth. So when we discuss how gender socialization affects men and women, also keep in mind that racism, ethnocentrism, ageism, homophobia, and transphobia exist alongside socialized gender roles and sexism. These are the intersectionalities we discussed in the previous chapter, and unfortunately, many people face multiple forms of discrimination.

Sometimes, gender socialization can have life-or-death consequences for babies. For example, in many nations, more boy than girl babies are carried to term and born. In South and East Asian nations like Vietnam and India but also in other areas of the world, such as the Caucasus and Southeast Europe, the sex ratio of births is more than 110 males for every 100 females, and in China, it is 115 males for every 100 females (UNFPA, 2012; Leopold, Ratcheva, & Zahidi, 2017). Recent headlines in the United Kingdom reveal that "sex-selective" abortions occur in highly developed nations as well (Watt & Newell, 2013). One societal consequence of this female infanticide is that there is now a shortage of women for men to marry in these nations. On an even deeper level, infanticide reveals that males are more valued than females.

As males and females move from birth to childhood to adulthood, they learn that they are expected to conform to the gender roles society has assigned to them. This process involves learning what toys to play with, how to speak (what tone, how often, to whom, etc.), how to present oneself in public, what sports to play (if any), what jobs or professions to consider, and so on. The authors of this book are the proud aunt and uncle of 21-year-old triplets (among a total of 13 amazing nieces and nephews)! Two of the triplets are boys and one a girl. It has been educational for them to watch the various gender socialization messages their niece has received, mostly from her peers and other socializing agents outside the family, and imagine how this might negatively affect her life chances and opportunities.

Gender socialization affects how men and women view the world as well as each other. For example, the political issues people deem most important tend to vary by gender. There is much evidence that women today, as a group, have different political opinions from men. There has been a "gender gap" in every U.S. presidential election since 1980. The gender gap refers to the percentage difference in men's and women's choice for president. For example, in 2012, 55 percent of women but only 44 percent of men voted to reelect Barack Obama to the office of U.S. president (The Washington Post, 2012). In 2016, the gender gap continued with only 42 percent of women but 53 percent of men voting for President Trump; notably, the racial gender gap was even higher, with 94 percent of black women and 68 percent of Latino or Hispanic women voting for Hilary Clinton while 53 percent of white women voted for Trump (Huang, Jacoby, Strickland, & Lai, 2016; Rogers, 2016).

Gender Socialization and Sex Roles

Gender socialization also includes learning sex roles (how to act in dating situations and sexually). As they grow older, boys and girls learn different lessons about with whom it is appropriate for them to act in a sexual manner, when it is permissible to have sex, and how they should have sexual relations. Although some societies are more open than others and many are becoming much more accepting of same-sex romantic relationships, the majority of people in almost all modern societies view heterosexual behavior (sex between men and women) as more appropriate and socially desirable than homosexual or bisexual behavior. However, a major paradigm shift has occurred in the United States. In 2001, only 35 percent of

Americans supported same-sex marriage while 57 percent opposed it; by 2017, 62 percent of Americans supported same-sex marriage with only 32 percent opposing it (Pew Research Center, 2017).

Perhaps some of this shift can be attributed to television shows with gay characters. Shows such as *Modern Family, Glee, Teen Wolf, Grey's Anatomy, How To Get Away With Murder, Six Feet Under, The Fosters, Brooklyn Nine-Nine, Will and Grace,* and many others have brought gay people into the living rooms of millions of Americans and helped change many people's attitudes about gay men, women, and youth.

The music industry has also begun to recognize and support more gay artists. One notable recent example of these changes comes from the world of hip-hop. Frank Ocean, a well-known rhythm and blues and hip-hop artist, came out as bisexual in 2012 and was immediately supported by renowned figures in the field such as Russell Simmons and Jay Z. Even Tyler, the Creator, who has penned many antigay lyrics, has publicly supported Ocean, with whom he has collaborated in the group Odd Future (McKinley, 2012). In 2013, Macklemore hit the top of the charts with his song "Same Love," painting gay marriage as a human rights issue and scolding those in the hip-hop community for their homophobia. Madonna joined him when he performed his song at the Grammy Awards. Immediately after the song, Queen Latifah conducted a live marriage ceremony for thirty-three couples, including many same-sex couples, and the Out Music Awards were established in 2001 to honor LGBTQ artists (see http://www.outmusicawards.com/).

Other examples are two National Football League players, Brendon Ayanbadejo and Chris Kluwe, who have publicly proclaimed their support for gay marriage (Himmelsbach, 2012). Michael Sam announced that he is gay after being drafted by the NFL, and Ryan O'Callaghan, a gay athlete, played six seasons. Esera Tuaolo, who declared he was gay while still playing and then competed on the Voice in 2017, wrote the book *Alone in the Trenches: My Life As a Gay Man in the NFL* (Tuaolo & Rosengren, 2007) and started up the anti-bullying organization Hate is Wrong (see www.hateiswrong.com). Jason Collins was the first openly gay player in the National Basketball Association (NBA). As an indication of the changing support for gay athletes, Collins's NBA shirt has become one of the most in demand in the whole league, despite the fact that he is not one of the game's superstars (Net Income, 2014).

Legal support for same-sex marriage is also growing across many areas of the globe. Since The Netherlands recognized same-sex marriages in 2000, other nations such as Belgium, Canada, Malta, Spain, South Africa, Germany, Norway, Sweden, Finland, Portugal, Iceland, Mexico, Argentina, Denmark, Uruguay, Colombia, Austria, New Zealand, France, Luxembourg, Ireland, Brazil, Australia, Netherlands, Britain, and the United States have also officially accepted such marriages, either nationwide or by jurisdiction.

In 2015, the Supreme Court ruled that same sex marriage is legal nationwide. On the flip side, Uganda has cracked down legally on homosexuality, and those who are identified as gay are often met with deep discrimination and violence.

The battle to end discrimination against lesbian, gay, bisexual, or transgender (LGBT) people in the United States is far from over, however. The majority of states still do not protect gay workers from discrimination. As of May 2014, it was still legal in 29 states to fire someone because of his or her sexual orientation (Confessore & Peters, 2014). Anti-transgender bills exist in Iowa, Kentucky, Missouri, Oklahoma, South Dakota, Tennessee, Colorado, New Hampshire, Pennsylvania, Indiana, West Virginia, Georgia, and Kansas. According to the ACLU, these bills discriminate against transgender individuals by "barring access to or even criminalizing the use of appropriate facilities, including restrooms, restricting transgender students' ability to fully participate in school, authorizing healthcare discrimination against trans people, allowing religiously-motivated discrimination against trans people, or making it more difficult for trans people to get identification documents with their name and gender" (ACLU, 2019).

Moreover, hate crimes continue to be committed against those perceived to be outside the heterosexual norm. In 2015, 85 percent of high school LGBTQ students faced verbal harassment because of their sexual orientation; 58 percent felt unsafe at school because of their sexual orientation; 43 percent felt unsafe because of their gender identity; 27 percent were physically harassed because of their sexual orientation; and 13 percent were physically harassed because of their gender identity (Williams Institute, 2017). For these reasons, LGBTQ teens are 3.5 times more likely to attempt suicide than their heterosexual peers, and transgender youth are almost 6 times more likely (Carroll, 2018).

Gender Roles

Gender roles basically set up binary gender expectations for males and females and confine us to certain types of behavior, limiting our freedom to act otherwise without fear of social disapproval. Those who vary from their gender-based roles or who do not fit binary expectations face negative sanctions. For example, husbands who choose to stay at home and take care of their children while their wives work outside the home may still be looked upon as not "true" men. Several years ago, the authors had a friend who was called a "girly man" because he took time off from full-time employment to be at home with his young children while his wife worked.[4] However, his wife faced no such negative sanctions when she stayed home with their children for the preceding eight years!

Today, though, gender roles for mothers and fathers have started to shift. Because of the growing need for two incomes to support a family, there are fewer stay-at-home mothers. Among couples who decide that one parent should stay at home, it is no longer an automatic choice for the stay-at-home parent to be the mother. While in 1975, only 47.4 percent of mothers with children under 18 worked outside the home, 71.1 percent of such mothers did so in 2017 (Bureau of Labor Statistics, 2018). Most couples decide that if they can afford to keep one parent at home, it should be the parent who would make the least money—and in four out of five households with preschool children, the mother is the primary caregiver (U.S. Bureau of the Census, 2011). In most states, even basic day care at a child care center costs more than tuition at a public university. The burden can be overwhelming for parents from the lower to upper middle classes. Some women decide that the cost is too much and drop out of the workforce to care for their child. Many of those who do so tend to have trouble reentering the workforce and attaining a job with similar status and pay.

Fathers who stay in the paid workforce also face the strain of dealing with changing gender roles. According to a 2011 study (National Study of the Changing Workforce), men now feel that they must be both providers and active parents and partners. Today, men believe that an "ideal" man "is not only a good employee, working long hours to be a successful breadwinner, but also an involved and nurturing husband/partner, father and son" (Aumann, Galinsky, & Matos, 2011, p. 3). Just as working mothers have faced the strain of trying to "do it all," working fathers are now feeling the pressure to meet expectations from both their family and their employer (and their own perceptions of an "ideal" man).

Gender Roles and Power

Despite the recent changes in gender roles that have accompanied women's increased role in the economic sphere, the behaviors we assign to masculine and feminine roles still provide men with more social, political, and economic power than women. Likewise, cisgender people are advantaged in society over transgender and non-binary individuals. In general, although the norms for gender roles are constantly evolving, boys are still more often trained to be tough, competitive, and self-promoting,[5] whereas girls still tend to be socialized to be sensitive, cooperative, caring, and self-deprecating. These respective gender roles clearly give cisgender men an advantage in both private and public arenas (Commuri & Gentry, 2005; Slaughter, 2012).

Even in societies where there is *de jure* (legal) equality for women, gender socialization can promote *de facto* (in fact) inequality between the two traditional sexes. For example, in the United States, there are more women than men. However, women have less power than men, as non-binary people have less power than binary people, and this leads to social, political, and economic inequalities between men and women in the United States. As you will see, these different aspects of power relations are interrelated, and they feed off one another.

Social Inequalities

Through gender socialization, both sexes are more likely to view men rather than women as experts and in possession of the tools of legitimate power. The different ways men and women communicate, with men more likely to interrupt and speak over others, influence the power dynamics between the sexes (Tannen, 2001; Wood, 2009). The fact that most people in leadership positions are men reinforces the perception that men have more leadership abilities and should be paid attention to more than women (Ely, Ibarra, & Kolb, 2011). This pattern appears throughout society. The traditionally male-dominated Harvard Business School and Fortune 500 Companies (of which only 24 have female CEOs) are two notable examples (Mejia, 2018). This attitude often manifests itself in everyday interactions when men and women enter a public event, meeting, or place of business together. Men tend to be greeted and attended to first.

The tasks that men and women are socialized to do in their own households also benefit men over women. As the number of dual-income households has increased, in the majority of households, women are still expected to come home from work and shoulder the majority of the burden of domestic chores, including cooking, cleaning, and especially child care. Arlie Hochschild (2003) has famously called this social phenomenon "the second shift."

Today, working women who are married spend less time on housework than women used to in earlier decades, when they tended not to work outside the home (University of Michigan Institute for Social Research, 2007). However, one out of two women but only one out of five men do household chores, such as laundry and cleaning, on a typical day. Many more men have started to cook or clean up after meals, but women are more likely to do these chores as well (43 percent of men compared with 70 percent of women; Bureau of Labor Statistics, 2016). These figures make it clear that women still bear the brunt of the second shift.

In addition to giving up more of their time for housework, women are also more likely than men to give up their own spending money for the good of the family. Men are more likely to spend money on themselves (Commuri & Gentry, 2005; Zeliker, 2011). These findings have made lenders and policymakers realize that families—and whole societies—tend to benefit more when women have control over the family finances (Zeliker, 2011). Internationally, this social fact is the main reason why most microcredit and microfinancing loans are granted to women, as they are much more likely to use the money responsibly for the advancement of their families.

Political Inequalities

Although the first wave of the women's movement led to the passage in 1920 of the Nineteenth Amendment, which gave women the right to vote, women are still far from attaining equal representation in politics. Today, more women vote than men. However, as of 2019, women number only 9 out of 50 state governors and only 102 out of 435 representatives and 25 out of 100 senators in the U.S. Congress—and this after a record number of women were voted in during the 2018 midterm elections (Center for American Women and Politics, 2019). Globally, only 15 out of 143 current heads of state or government are female (Geiger & Kent, 2017).[6]

Women who run for political office (or any position of power) must, in some ways, resocialize themselves. They have to learn to be self-promoters rather than self-deprecators, to be tough as well as sensitive, and to speak up rather than defer to others. At the same time, they must trust that voters will be more impressed with their leadership credentials than turned off by their stepping out of the traditional "feminine" gender role. As anyone who followed the presidential race in 2016 knows, this juggling act proved difficult for Hillary Clinton (as it would for any woman running for president). For example, while many Americans were thrilled to vote for a woman for president, others frowned on Clinton's stepping outside the gender norm. While the United States has not yet elected a female president or even a female vice president, women have achieved greater political gains lower down the political ladder. For example, in 2019, 27.6 percent of state legislators were women, representing, according to CAWP (2019), a quintuple increase since 1971. Although the gain in the percentage of female state legislators has increased in recent years, women are still vastly underrepresented, which is very important because many officials in higher office start their political careers in these types of political positions.

Economic Inequalities

Women have made tremendous progress in the economic sphere over the past five decades. The civil rights movement and the second wave of the women's movement brought about the passage of the Equal Pay Act in 1963 and the Civil Rights Act in 1964, Title IX's prohibition of sex discrimination in schools (and in funding for school sports), acknowledgment of sexual harassment as a social issue, and the inclusion of women in affirmative action programs. These legislative acts, inflation, and the decline in the number of jobs paying enough for one earner to support a family all contributed to women making up 47 percent of the workforce in 2017 (Bureau of Labor Statistics, 2019). However, in 2017, women held only 10.6 percent of board seats and 4.8 percent of CEO positions at Fortune 500 companies (Catalyst, 2018; Zarya, 2018).

While there are fewer women than men on the top of the economic ladder, women make up a disproportionate percentage of those clinging to the bottom of the ladder. Women who are heads of households and their children are particularly likely to live in poverty. In 2016, 5.1 percent of families led by a married couple lived in poverty. However, 26.6 percent of families headed by a single mother lived in poverty, compared with 13.1 percent of families headed by a single father (Semega, Fontenot, & Kollar, 2017). Overall, nearly one in five (19 percent) U.S. children lives in poverty (Koball & Jiang, 2018).

Education also plays a role in who marries and the resulting inequalities. Twelve percent of mothers in their twenties with a college degree give birth outside marriage, compared with 59 percent of mothers with a high school degree or less (Child Trends, 2014). The divorce rate for more educated parents is also lower than for those with less education (Heller, 2012). The result is a growing gap in income and other parenting resources between married parents and single parents. This gap, in turn, creates a disparity in life chances between the children of these different types of parents. This inequality tends to be passed down further and leads to a cycle of poverty, as those with less educated and single parents are more likely to leave school earlier and have their own children outside marriage (DeParle, 2012).

The Wage Gap

Despite all the progress that the women's movement has made, women today still make only 82 cents for every dollar made by men (Graf, Brown, & Patten, 2018) and still face a "glass ceiling" when trying to advance into higher levels of management, for a variety of reasons including the following:

- Gender socialization influences (a) what subjects girls study and are encouraged to study in school (e.g., girls are still heavily underrepresented among math and science majors in college),

(b) what career paths they are directed toward and choose, and (c) their ability to think of themselves, present themselves, and be perceived by others as capable professionals and leaders.

- Only women can become pregnant and give birth to children. If women want to have children, they must take at least some time off work to give birth, recover from childbirth, and care for their newborn. Women in heterosexual marriages end up doing most of the child care and housework even when working outside the home (Hochschild, 2003). This affects their ability to spend extra hours at their paid job and advance in their career. Ironically, this factor further fuels the stereotype that women are less capable than men and less hardworking, when in fact they are juggling more and working longer hours than men (when housework, child care, and professional work are combined).

- Sex discrimination and sexual harassment still exist and work to hamper the economic progress of women, as the following examples show:

  o A 2012 Yale study revealed that both female and male scientists ranked identically qualified candidates for a lab manager position differently based on their gender. The male candidates were perceived as more competent and deserving of higher salaries than the female applicants (Moss-Racusin, Dovidio, Brescoll, Graham, & Handelsman, 2012).

  o New York City settled a gender bias lawsuit against the Emergency Medical Service of the New York City Fire Department for more than $1.25 million. The plaintiffs maintained that it was much harder for women than men to earn promotions above the rank of lieutenant (promotions to lieutenant were based on Civil Service exam results) (Goldstein, 2013).

  o Bank of America paid $39 million in a settlement decision for discriminating against women in its Merrill Lynch brokerage operation (McGeehan, 2013).

  o The U.S. Equal Employment Opportunity Commission forced FedEx Freight to pay $115,000 for choosing an unqualified man for a position over three qualified women, Amtrak to pay $171,483 to compensate a female worker for wage discrimination and then retaliating against her for complaining, and the Big Vanilla Athletic Club to pay $161,000 for sexually harassing female employees and then firing them when they complained (Equal Employment Opportunity Commission, 2009, 2011, 2012).

Clearly, challenges still exist for women in the social, political, and economic spheres. However, progress toward equality has been made through effective, organized efforts. Our culture and gender roles are continually changing and adjusting to the social and structural forces at work in our society. Legislation like the Nineteenth Amendment, the Equal Pay Act of 1963, and the Civil Rights Act of 1964, which prohibited employment discrimination on the basis of sex (as well as race, color, religion, and national origin), did much to change the actions and, gradually, gender socialization of Americans. Sociological tools have been and will continue to be effective means for uncovering and addressing gender inequality in U.S. society and across the world. The Sociologist in Action section below is a good example of how sociology can be effectively used in this endeavor.

Sociologist in Action

Melissa Sheridan Embser-Herbert

During the summer of 2011, I had the opportunity to make my first visit to Rwanda, explore the culture, plan a course bringing students to Rwanda, and learn, more directly than books allow, about the 1994 genocide. Yes, I saw the mountain gorillas—up close and personal—and I bought too many hand-woven baskets, but I also saw—and smelled—stacks of bones and piles of bloodied clothing at genocide memorials. In 2012, with a colleague and 12 students, I returned to Rwanda to study restorative justice and reconciliation in postgenocide Rwanda.

It's one thing to watch a video or read a book about genocide. It's another to visit the country where it happened and see the mass graves, the number of people missing limbs, and—as my colleague experienced—talk with someone who says, "I killed seven people. I've made peace with six families. I hope to with family seven." While our central aim was to learn about restorative justice and reconciliation, we also considered other issues that impact postgenocide Rwandan society. One key issue is the role of women in Rwandan society.

While we were in Rwanda, we visited the market in Musanze. In Rwanda, most women are employed in agriculture, with "off-farm" opportunities being quite limited. Women are far less likely to have access to financial services, including microcredit, often required for starting their own businesses. If the independent participation of women in the nonagricultural labor market is to increase, more support must be found for women who have the skills and

(Continued)

(Continued)

initiative to become entrepreneurs. Via our interpreter, one student, Angela, started talking with one of the women at the market who sewed, wanting to know about her life. On the class blog, Angela wrote, "Mama Joyce went on to tell me that she makes 30,000 RWF (Rwandan francs), or about 50 USD [US dollars], per month, and from that she pays 20,000 RWF, or 33 USD, in rent for her sewing machine. . . . Her income supports the majority of her family's basic needs and definitely doesn't allow for anything extra. They net only 17 USD a month. I was astounded to learn that the cost to purchase her own machine would be about 70,000 RWF, or 115 USD. Cripes! I spend approximately that much on my cell phone bill each month." Angela decided to make a personal donation so that Mama Joyce could purchase a sewing machine. But she had also received $100 from her department in case she had the opportunity to conduct interviews while in Rwanda. Angela was told that should that not be possible, she could donate the money as she wished. She decided to give Mama Joyce that money for the purchase of a second sewing machine. The idea was that the women who sewed together might start a cooperative, a business that they would run together, sharing in its profits. But she didn't stop there. Upon her return to the United States, Angela started exploring how to establish her own nonprofit to help other women in Musanze, Rwanda. *Begin* was begun.

Begin's mission was to raise funds to allow more women in Musanze to purchase their own sewing machines. Referring to the money that would no longer be spent renting machines, Angela writes, "This is money that she [Mama Joyce] could be using to buy food and clothes for her children . . . or money she could be using to reinvest in her community. . . . By building a park, for example. Or maybe a kindergarten." Mama Joyce is the primary wage earner for her family. Her husband, having left military service, has been unable to find permanent, full-time work. This shift in responsibility is not unique to Mama Joyce, or to Rwanda. Women across the globe share these experiences. Angela recognized this need and then took steps to address it.

My students and I used our sociological imaginations to recognize patterns of systemic inequality in the institutions of Rwanda, including the economy. Our commitment to using that knowledge to make a positive impact on society led some to take action. While these steps are relatively small, we know that there are many other sociologists in action—some reading this piece right now—taking steps of their own. Together, we can change the world for the better.

Source: Courtesy of Melissa Sheridan Embser-Herbert.

Exercise 9.1
It Gets Better

1. Read at least one of the research reports at the Gay, Lesbian, and Straight Education Network (http://www.glsen.org/cgi-bin/iowa/all/research/index.html).

2. Go to the It Gets Better Project website (http://www.youtube.com/user/itgetsbetterproject) and watch at least two of the It Gets Better videos, and then watch Laverne Cox (https://www.youtube.com/watch?v=1Mfx tM9N3fw&list=PL8EC60EF703CBD74C).

3. Then, go to https://vimeo.com/76299916, and watch Jonah Mowry's "Whats Goin On."

4. Write a two- to three-page reflection paper that describes
   a. some of the findings from the research report(s),
   b. how those findings relate to the videos you watched,
   c. how your own sexual orientation (and related experiences) influence your reaction to the videos and report(s), and
   d. what steps you can take to make society a safer place for gays and lesbians.

Exercise 9.2
Gender, Relationships, and Socialization

1. Make a list of the 10 characteristics you are most looking for in a sexual or life partner. Be sure to include your gender (but not your name) at the top of the page (female, male, third gender/non-binary, or other).

2. On a separate piece of paper, make a list of the 10 characteristics that you think a prospective partner is most looking for in you. Be sure to include your gender (but not your name) at the top of the page (female, male, third gender/non-binary, or other).

(Continued)

(Continued)

3. Have your professor collect these and tally up the results, following the instructions we have provided on this book's website (http://study.sagepub.com/white6e).

4. Analyze the results. What are men most looking for? What are women most looking for? What are third gender/non-binary individuals most looking for? What are the differences between these? What are the causes of these differences? What are the social effects of these differences—that is, how do they affect the gender roles and the social power of men, women, and third gender individuals?

5. For further analysis, explore the following: What did women think that potential partners were looking for in them? What did men think that potential partners were looking for in them? What did non-binary individuals think that potential partners were looking for in them? Compare these results with those found in the responses to Question 4.

6. Were women, men, or non-binary individuals more accurate in guessing what their potential partners were looking for? Why? How does gender socialization help explain this?

7. How does gender socialization promote heterosexuality at the expense of other sexual orientations?

8. Read and answer "The Heterosexual Questionnaire" at http://higherlogicdownload.s3.amazonaws.com/NASN/e277e492-64b1-4f55-ac15-00857a7a5662/UploadedImages/Oregon%20Microsite/Documents/HeterosexualQuestionnaire.pdf and now go back and answer Question 7. Have your answers changed or have you added to them in any way? Why or why not?

Exercise 9.3
Discrimination and You

1. Make a list of five occasions when you have been discriminated against in your life. At least three of these should be based on some

biological fact that you can't change about yourself (your race, height, sex, age, sexuality, hair color, body type, etc.). Try, if you can, to list multiple biological facts for which you have been discriminated against.

2. Make a list of five occasions when you have been prejudiced against or discriminated against someone (or a group of people). Make sure that at least three of the five were based on biological facts that the person (or group of people) couldn't change, such as race, sex, sexuality, or physical appearance.

3. Write a two- to three-page essay that discusses (a) your own experiences of being discriminated against, including how it felt and how it made you act and react; (b) why you were prejudiced or discriminated against others; and (c) how a person can face multiple points of discrimination.

4. Make sure to include at least one full paragraph addressing what you might be able to do, in your own life, to be less prejudiced and less likely to discriminate against other groups of people.

Exercise 9.4
Gender Differences in Personal Ads

1. Write an advertisement about yourself for an online dating site. In your ad, describe yourself (in at least 40 words) and what you are looking for in a partner (in at least 40 words). Be honest and realistic in what you write, as if this truly were an ad you would submit to find a romantic partner.

2. Get in groups of four (preferably with some gender variation), and read your ads to each other. What do these ads tell you about gender and the way we are socialized to maintain and conform to gender roles? Dig beneath the surface in analyzing (a) what is included in the ads and how this reflects gender socialization and (b) what isn't included in the ads and how this reflects gender socialization.

Exercise 9.5
Media Images of Women and Men

Women, read the following statement by Lynne Grefe, CEO of the National Eating Disorders Association, and answer the questions below it.

> In 1965, the average fashion model weighed just eight percent less than the average American woman. . . . The average fashion model today is 5'11" and weighs 117 pounds, which makes her thinner than 98 percent of women. Three-quarters of the female characters in TV situation comedies are underweight. Yet the average woman stands 5'4", weighs 140 pounds and wears between a size 12 [and] 16. We know that continued exposure to unrealistic images is linked to depression, loss of self-esteem and the development of unhealthy eating habits in women and girls. Genetics may load the gun, but society pulls the trigger. (National Eating Disorders Association, 2009, para. 5)

1. Do you think the media regularly exposes you to unrealistic images of women? Why or why not?
2. Do you ever find yourself trying to mold your body into one of those unrealistic images? Why or why not?
3. What do you think you and other people can do to (a) counter the effects of and (b) diminish the prevalence of such unrealistic portrayals of women in the media?

Men, read "If Being a Man Means Having Body Hair and Sweating, Why Are the Sexy Guys in Ads Immune to Both?" at http://seedmagazine.com/content/article/the_media_assault_on_male_body_image and answer the following questions:

1. Do you think the media regularly exposes you to unrealistic images of men, such as those described in the article? Why or why not?
2. What is your reaction to the findings described in the article? Does your answer relate to the amount of media you tend to consume? Why or why not?
3. What do you think you and other people can do to (a) counter the effects of and (b) diminish the prevalence of such unrealistic portrayals of men in the media?

Exercise 9.6
A (Gendered) Tour Through a Toy Store

Go to a major (chain) toy store, and try to find a toy (not a piece of sports equipment) that is not marketed specifically toward a boy or a girl. Be sure to take note of the following:

1. What is the layout of the store? Are there distinct boys' and girls' sections?

2. What types of toys are marketed toward girls? What types of toys are marketed toward boys?

3. What types of words and descriptions are used on the toys to market them toward girls or boys?

4. Are there many toys that do not indicate on the packaging (e.g., have pictures of either all girls or all boys on it) the sex of the child for which they are deemed appropriate?

5. How did you find the toy you selected? Was it easy or difficult to find? Where was it located? Based on your experience in the toy store, write a two- to three-page paper that analyzes how toys are a means of gender socialization. What values do they teach boys and girls? Why? For what social roles are these toys teaching and training boys and girls?

Exercise 9.7
The Storm

The following videos provide good illustrations of the ongoing struggle over cultural norms about same-sex relationships: "A Gathering Storm" at https://www.youtube.com/watch?v=OOoXRBzHgds and the parody of it, "A Gaythering Storm" (https://www.youtube.com/watch?v=PhaihBGD2tc).

(Continued)

(Continued)

Watch the videos and answer the following questions:

1. Did you find yourself agreeing or disagreeing with the message of "A Gathering Storm"? What about your gender socialization led you to feel this way? Did your religious socialization contribute?

2. How did you react to "A Gaythering Storm"? What about your gender socialization led you to feel this way?

3. Fifty years from now, how do you think most Americans will view each of these videos? Why? Does your response provide evidence that gender roles are social constructions? Why or why not?

Exercise 9.8
Education Is the Key to Success! (More success for men, though)

1. Read the Executive Summary of the report "Women Can't Win" (https://1gyhoq479ufd3yna29x7ubjn-wpengine.netdna-ssl.com/wp-content/uploads/Women_ES_Web.pdf).

2. Write a two- to three-page paper. What did you learn? Were you surprised? What data stuck out most to you? What does this say about sexism, our society, socialization? How would Cooley analyze this report through his concept of the looking-glass self, and how would Jane Addams analyze the report?

Exercise 9.9
Gender, Politics, and Political Power

Go to the CAWP website (http://www.cawp.rutgers.edu/). Browse through the information on the site, and read some of the research it contains on women and politics.

1. What did you learn about women and the political realm? Does the information you've uncovered surprise you at all? Why or why not?

2. Have you ever thought about running for political office? Why or why not?

3. (Option 1) If you have thought of running, do you think your gender would affect the likelihood of your attaining the political office you sought? Based on the information found on the CAWP website, discuss how you think your gender would (a) affect how you would be portrayed by the media, (b) how the public would perceive you, and (c) your chances of winning.

   (Option 2) If you haven't thought of running, do you think gender socialization has affected this decision? Why or why not? Pretend now that you are interested in running, and answer the questions posed in Option 1.

4. Imagine that you are the campaign manager for a female candidate running for president. How would you advise her? What should she stress in her campaign speeches? How should she speak? How should she dress? What issues would you suggest she take on, and what issues would you advise her to stay away from?

5. How might your advice be different if the candidate were a man? What concerns might you have that you wouldn't have if the candidate were a woman? What might you not have to worry about if the candidate were a man instead of a woman?

6. What do your answers to Question 5 tell you about the importance of gender roles in the public imagination? What do they tell you about the social construction of these roles? What do they tell you about the connections between gender and power?

Exercise 9.10
Leadership Qualities

This exercise will require three short surveys to be administered in sequence to three different pools of respondents. (The easiest method would be to

(Continued)

(Continued)

administer the survey to three of your classes, making sure no student responds to more than one of the surveys.)

1. Create a short form with one question and space for about 10 answers. Ask the respondents to write down their gender (but not their names) on the top of the page. Invite non-binary categories, as Human Rights Campaign suggests, including female, male, and third gender/non-binary. The question on the form should read, "What personal qualities make one a natural leader?" Ask the respondents to write a list of about 10 qualities that make one a natural leader. Collect answers from a class that has at least 20 students.

2. Compile a list of all the answers given by the respondents. Then, order them from the most commonly offered responses to the least common. Finally, select the top 10 responses in terms of frequency.

3. Create a form with a four-column table, and list the top 10 personal qualities from the first survey under the heading "Personal Attributes." Label the other columns "Masculine," "Feminine," and "Equally Masculine & Feminine," as shown on the next page.

Personal Attributes	Masculine	Feminine	Equally Masculine & Feminine
Attribute 1			
Attribute 2			

4. Ask the members of a different class (with at least 20 students) to rate each attribute as fitting best into one of the three categories. Ask the respondents to write down their gender (but not their names) at the top of the page. Collect the forms when they are finished.

5. Summarize how many of the popular leadership qualities are associated with masculine identity and how many are associated with feminine identity.

6. Survey a different class (with at least 20 students), using a new form that lists all of the attributes and asks respondents to indicate their gender and to evaluate each attribute on a scale from least to most desirable, as illustrated in the table that follows.

For each of the following personal attributes, please circle the number that indicates how desirable you find it as a feature of a woman's personality:

Personal Attributes	Most Undesirable	Neutral	Most Desirable
Attribute 1	1 2 3 4	5 6	7 8 9 10
Attribute 2	1 2 3 4	5 6	7 8 9 10

7. Collect the forms and average the ratings. Overall, how desirable did the respondents find leadership qualities to be in women?

8. Modify the form in Step 6 to measure how desirable those characteristics are in men. Compare the results.

9. Compare the responses of male, female, and third-gender respondents and write a two- to three-page paper summarizing and sociologically analyzing your results. What do you believe are the causes and effects of the trends you have identified in your data?

Exercise 9.11
Where Is It Illegal to Be Gay?

Read "74 Countries Where Homosexuality Is Illegal" at https://76crimes.com/76-countries-where-homosexuality-is-illegal/ and write a two-page paper outlining what you have learned, highlighting at least two countries as you also take a look at the bigger picture globally.

Exercise 9.12
Promoting Gender
Equality Throughout the World

Go to the United Nations Development Funds' *State of World Population 2013: Motherhood in Childhood* at https://www.unfpa.org/publications/state-world-population-2013 and watch at least two of the short videos on the site.
Then, answer the following questions:

1. What stories were told in each of the videos?

2. How did the videos reveal the relationship among education, adolescent motherhood, and gender equality?

3. How did they influence your views on gender inequality throughout the globe and the most effective means to address it? Why?

Extra credit: Write a one- to two-page paper that uses your sociological imagination in developing the plan for a program that would promote gender equality worldwide (or in a specific country). Be creative, strategic, and specific.

DISCUSSION QUESTIONS

1. If you grew up in a cisgender household, think about the division of labor along gender lines in your house. If you grew up in a household that wasn't cisgender, think about the division of labor and if one person filled more traditionally "feminine" roles and if one person filled more traditionally "masculine" roles. Who did what? Why was it like that? Would you want to maintain or change this in your own adult household? Why? If you are cohabitating now (with a spouse, partner, or roommates), discuss how you divide the labor in your current household and if it is similar to the way household chores were distributed in your family of origin (and why or why not).

2. What does "equal rights" mean to you? How much equality is there between genders in U.S. society? How much equality is there between

cisgender and transgender or non-binary individuals? How much equality is there between heterosexual and homosexual or bisexual individuals?

3. Gender role socialization begins at birth or before. Friends and relatives want to know the sex of the expected child so that they can purchase gender-appropriate blankets, pillows, bibs, and clothes (blue or pink). The socialization continues and becomes more intense as children get older. Is it possible to raise boys and girls in the same way? Is it desirable? What are some of the obvious ways in which children are taught gender roles? Can people change those to reduce gender-based discrimination? Should we? If so, how? Be specific in outlining some proactive ways to create gender equity through changing gender roles and expectations.

4. How are gender roles related to sex roles? What do they teach us about how to act sexually and with whom to have sex?

5. Why should men be interested in promoting equal rights and opportunities for women? What are some appropriate ways for men to help women gain greater social, political, and economic power?

6. Why should cisgender people be interested in promoting equal rights and opportunities for transgender people? Why should heterosexual people be interested in promoting equal rights and opportunities for LGBTQ people? What are some appropriate ways in which people from the dominant groups in society can act as allies to those who are not?

7. Would you feel (a) more comfortable and (b) more secure with a female or a male president? Why? How do you think gender socialization influenced your answer? How do you think the courses that made you aware of gender socialization influenced your answer? If you answered that you would feel more secure with a man as president, what would need to happen to change your mind?

8. What do you think would be the most effective way to diminish social inequality between genders in the United States? Political inequality? Economic inequality?

9. If you were to conduct a study about sex discrimination, how would you go about doing so? What theoretical perspective would best help you make sense of your findings?

10. Describe how women of color are "doubly oppressed" and lesbian or transgender women of color are "triply oppressed" in our society. Does your friendship circle include people of various (a) genders, (b) sexual orientations, and (c) races? Why or why not?

SUGGESTIONS FOR SPECIFIC ACTIONS

1. Are you interested in promoting economic equality for women? Go to the website of 9 to 5, National Organization of Working Women, at http://www.9to5.org. Read about the efforts 9 to 5 is making to fight for the issues of working women. If you want to become part of one of the 9 to 5 campaigns, follow the steps on their "Get Involved" and "Action Alert" pages.

2. Go to the Human Rights Watch website at http://hrw.org, learn about their campaigns to stop human rights violations against women, and participate in one (write "women" in the search bar).

3. Go to the Human Rights Campaign action page at https://act.hrc.org/ea-campaign/action.retrievestaticpage.do?ea_static_page_id=5614 and learn about the various actions you can take to support LGBTQ rights. Participate in one, and organize a campaign on campus to encourage others to participate in it as well.

   Please go to this book's website at http://study.sagepub.com/white6e to find further civic engagement opportunities, resources, and peer-reviewed articles related to this chapter.

NOTES

1. A very small percentage of the population is born with both male and female sexual characteristics (known as intersexuality). Almost always in these cases, an operation is performed shortly after birth to make the child's sex organs primarily male or female (usually female).
2. Sex change surgery is possible, though it may be difficult and expensive. An increasing number of people are choosing to pursue this surgery, though opportunity may be limited by financial means and societal stigma.
3. While hormone therapy and sex change surgery are important steps for some transgender people, it is difficult to access due to the expense, variable health insurance coverage, and other barriers to access in healthcare.
4. His comments about making conversations with the other parents (all mothers!) waiting to pick up their children from school or soccer are quite amusing. While he can hold his own when the conversations turn to laundry, he says he would much rather be discussing sports or politics and that he feels like an outsider with he considers the "gossip culture" of the mothers. Just because people have the power to defy some gender role expectations does not mean that gender socialization has not affected them at all.
5. And boys are taught to distance themselves from "girlish" behavior. The saying "You cry like a girl" is one of the many ways boys are taught to be anything but "like a girl."

6. See the Center for American Women and Politics (CAWP) website at http://www.cawp.rutgers.edu/index.php for more information about women in the political realm.

REFERENCES

American Civil Liberties Union. (2019). Legislation affecting LGBT rights across the country. Retrieved from https://www.aclu.org/legislation-affecting-lgbt-rights-across-country

American Sociological Association. (2016, May 24). Sociologists available to discuss transgender-related issues [Press release]. Retrieved from http://www.asanet.org/press-center/press-releases/sociologists-available-discuss-transgender-related-issues

Aumann, K., Galinsky, E., & Matos, K. (2011, July). *The new male mystique* (Families and Work Institute, National Study of the Changing Workforce). Retrieved from http://www.familiesandwork.org/site/research/reports/newmalemystique.pdf

Bureau of Labor Statistics. (2016, December 20). American time use survey. Retrieved from https://www.bls.gov/tus/charts/household.htm

Bureau of Labor Statistics. (2018, April 19). Employment characteristics of families summary. Retrieved from https://www.bls.gov/news.release/famee.nr0.htm

Bureau of Labor Statistics. (2019, January 18). Labor force statistics from the Current Population Survey. Retrieved from https://www.bls.gov/cps/cpsaat11.htm

Carroll, L. (2018, October 8). *LGBT* youth at higher risk for suicide attempts. Retrieved from https://www.reuters.com/article/us-health-lgbt-teen-suicide/lgbt-youth-at-higher-risk-for-suicide-attempts-idUSKCN1MI1SL

Catalyst, Inc. (2018). Women's representation on Fortune 500 boards inches upward. https://www.catalyst.org/media/womens-representation-fortune-500-boards-inches-upward

Center for American Women and Politics. (2019). Women in elective office 2019. Retrieved from http://www.cawp.rutgers.edu/women-elective-office-2019

Child Trends. (2014). *Non-marital births: Educational differences*. Retrieved from http://www.childtrends.org/non-marital-births-educational-differences/

Coker, M. (2018, June 22). Saudi women can drive, but here's the real roadblock. Retrieved from https://www.nytimes.com/2018/06/22/world/middleeast/saudi-arabia-women-driving.html

Commuri, S., & Gentry, J. W. (2005, September). Resource allocation in households with women as chief wage earners. *Journal of Consumer Research, 32*, 185–195.

Confessore, N., & Peters, J. W. (2014, April 28). Gay rights push shifts its focus to south and west. *The New York Times*. Retrieved from http://www.nytimes.com/2014/04/28/us/politics/gay-rights-push-shifts-its-focus-south-and-west.html?hp

DeParle, J. (2012). Two classes, divided by "I do." *The New York Times*. Retrieved from http://www.nytimes.com/2012/07/15/us/two-classes-in-america-divided-by-i-do.html?ref=us&pagewanted=all

Equal Employment Opportunity Commission. (2009, February 26). *Big Vanilla Athletic Club to pay $161,000 to settle EEOC lawsuit for sexual harassment.* Retrieved from http://www.eeoc.gov/eeoc/newsroom/release/2-26-09.cfm

Equal Employment Opportunity Commission. (2011, November 10). *Amtrak to pay $171,483 to settle EEOC sex/wage discrimination suit.* Retrieved from http://www.eeoc.gov/eeoc/newsroom/release/11-10-11a.cfm

Equal Employment Opportunity Commission. (2012, June 25). *FedEx Freight to pay $115,000 to settle EEOC sex discrimination lawsuit.* Retrieved from http://www.eeoc.gov/eeoc/newsroom/release/6-25-12b.cfm

Ely, R. J., Ibarra, H., & Kolb, D. (2011, September). Taking gender into account: Theory and design for women's leadership development programs. *Academy of Management Learning & Education, 10*(3), 507–527.

Geiger, A., & Kent, L. (2017, March 8). Number of women leaders around the world has grown, but they're still a small group. Retrieved from http://www.pewresearch.org/fact-tank/2017/03/08/women-leaders-around-the-world/

Goldstein, J. (2013, June 14). 5 settle gender bias suit against fire department. *The New York Times*. Retrieved from http://www.nytimes.com/2013/06/15/nyregion/women-settle-gender-bias-lawsuit-against-fire-dept.html

Graf, N., Brown, A., & Patten, E. (2018, April 9). The narrowing, but persistent, gender gap in pay. Retrieved from http://www.pewresearch.org/fact-tank/2018/04/09/gender-pay-gap-facts/

Heller, K. (2012). The myth of the high rate of divorce. *Psych Central.* Retrieved from http://psychcentral.com/lib/2012/the-myth-of-the-high-rate-of-divorce

Himmelsbach, A. (2012, September 9). Players' support of gay marriage alters N.F.L. image. *The New York Times*. Retrieved from http://www.nytimes.com/2012/09/09/sports/football/players-support-of-gay-marriage-alters-nfl-image.html?_r=1

Hochschild, A. R. (with Machung, A.). (2003). *The second shift*. New York, NY: Penguin Books.

Huang, J., Jacoby, S., Strickland, M., & Lai, K. R. (2016, November 9). Election 2016: Exit polls. *The New York Times*. Retrieved from https://www.nytimes.com/interactive/2016/11/08/us/politics/election-exit-polls.html

Human Rights Watch. (2016, July 16). Boxed in: Women and Saudi Arabia's male guardianship system. Retrieved from https://www.hrw.org/report/2016/07/16/boxed/women-and-saudi-arabias-male-guardianship-system

Koball, H., & Jiang, Y. (2018, January). Basic facts about low-income children. Retrieved from http://www.nccp.org/publications/pub_1194.html

Leopold, T. A., Ratcheva, V., & Zahidi, S. (2017). *The global gender gap report 2017.* Retrieved from World Economic Forum website: http://www3.weforum.org/docs/WEF_GGGR_2017.pdf

McGeehan, P. (2013, September 6). Bank of America to pay $39 million in gender bias case. *The New York Times.* Retrieved from http://dealbook.nytimes.com/2013/09/06/bank-of-america-to-pay-39-million-in-gender-bias-case/?_r=0

McKinley, J. C., Jr. (2012, July 6). Hip-hop world gives gay singer support. *The New York Times.* Retrieved from http://www.nytimes.com/2012/07/07/arts/music/frank-ocean-draws-praise-for-declaring-his-homosexuality.html?_r=1

Mejia, Z. (2018, May 21). Just 24 female CEOs lead the companies on the 2018 Fortune 500—fewer than last year. Retrieved from https://www.cnbc.com/2018/05/21/2018s-fortune-500-companies-have-just-24-female-ceos.html

Moss-Racusin, C. A., Dovidio, J. F., Brescoll, V. L., Graham, M. J., & Handelsman, J. (2012, September 17). Science faculty's subtle gender biases favor male students. *Proceedings of the National Academy of Sciences of the United States of America, 108,* 3157–3162.

National Eating Disorders Association. (2009, February 26). *National Eating Disorders Association unveils powerful and provocative ad campaign.* Retrieved from http://www.nationaleatingdisorders.org/press-rom/press-releases/2009-press-releases/national-eating-disorder-association-unveils-powerful-provocative-ad-campaign

Net Income. (2014, February 25). Jason Collins no. 98 top-seller Tuesday. *Netsdaily.* Retrieved from http://www.netsdaily.com/2014/2/25/5445580/jason-collins-no-98-goes-on-sale-tuesday

Pew Research Center. (2017, June 26). Changing attitudes on gay marriage. Retrieved from http://www.pewforum.org/fact-sheet/changing-attitudes-on-gay-marriage/

Rochlin, M. (1977, January). The heterosexual questionnaire. Unpublished manuscript. Retrieved from http://higherlogicdownload.s3.amazonaws.com/NASN/e277e492-64b1-4f55-ac15-00857a7a5662/UploadedImages/Oregon%20Microsite/Documents/HeterosexualQuestionnaire.pdf

Semega, J. L., Fontenot, K. R., & Kollar, M. A. (2017). *Income and poverty in the United States: 2016.* (U.S. Census Bureau, Current Population Reports, P60-259). Washington, DC: U.S. Government Printing Office.

Sirgany, S., & Smith-Spark, L. (2018, June 24). Landmark day for Saudi women as driving ban ends. Retrieved from https://www.cnn.com/2018/06/23/middleeast/saudi-women-driving-ban-lifts-intl/index.html

Rogers, K. (2016, November 9). White women helped elect Donald Trump. *New York Times,* p. 9.

Slaughter, A.-M. (2012, July/August). Why women still can't have it all. *The Atlantic.* Retrieved from http://www.theatlantic.com/magazine/archive/2012/07/why-women-still-can-8217-t-have-it-all/9020

Tannen, D. (2001). *You just don't understand: Women and men in conversation.* New York, NY: HarperCollins.

Tuaolo, E., & Rosengren, J. (2007). *Alone in the trenches: My life as a gay man in the NFL.* Naperville, IL: Sourcebooks.

UNFPA. (2012, August). *Sex imbalances at birth: Current trends, consequences, and policy implications.* Retrieved from http://www.unfpa.org/webdav/site/global/shared/documents/publications/2012/Sex%20Imbalances%20at%20Birth.%20PDF%20UNFPA%20APRO%20publication%202012.pdf

University of Michigan Institute for Social Research. (2007, January). Time, money and who does the laundry. *Research Update, 4.* Retrieved from http://www.isr.umich.edu/home/news/research-update/2007-01.pdf

USA Hockey, Inc. (2019). *2017–2018 season final registration reports.* Retrieved from USA Hockey website at https://cdn3.sportngin.com/attachments/document/2210-1687681/2017-18_Final_Registration_Report.pdf

U.S. Bureau of the Census. (2011, December 5). *One-third of fathers with working wives regularly care for their children* (Census Bureau Reports). Retrieved from http://www.census.gov/newsroom/releases/archives/children/cb11-198.html

The Washington Post. (2012, November 7). Exit polls 2012: How votes are shifting. Retrieved from http://www.washingtonpost.com/wp-srv/special/politics/2012-exit-polls/index.html#United-States

Watt, H., & Newell, C. (2013, October 7). Law "does not prohibit" sex-selection abortions, DPP warns. *The Telegraph.* Retrieved from http://www.telegraph.co.uk/health/healthnews/10360386/Law-does-not-prohibit-sex-selection-abortions-DPP-warns.html

The Williams Institute. (2018, March 22). LGBT youth experiences discrimination, harassment, and bullying in school. Retrieved from https://williamsinstitute.law.ucla.edu/press/lgbt-youth-bullying-press-release/

Wood, J. (2009). *Gendered lives: Communication, gender, and culture* (8th ed.). Belmont, CA: Wadsworth.

Zarya, V. (2018, May 21). The share of female CEOs in the Fortune 500 dropped by 25% in 2018. Retrieved from http://fortune.com/2018/05/21/women-fortune-500-2018/

Zeliker, V. A. (2011, January 27). The gender of money. *The Wall Street Journal.* Retrieved from http://blogs.wsj.com/ideas-market/2011/01/27/the-gender-of-money

Zevallos, Z. (2019). Sociology of gender. Retrieved from https://othersociologist.com/sociology-of-gender/

CHAPTER 10

Social Institutions

Family and Economy

W hat do social institutions have to do with your life? A lot! The makeup of your family, the laws you must follow, your professional career, your schooling, and even whether or not you believe in a higher power (and, if so, what kind of higher power) are all based on the social institutions in your society. You begin your life among family and learn about the world through educational institutions (schools) and perhaps through religious institutions (including rituals surrounding birth, marriage, and death). Much of your education is about preparing for life within structured economic institutions (the labor market).[1] All the while, your public life—and even your private one—is moved and shaped by the workings of political institutions. If the institutions change, so do you. Think about the impact of the Twenty-Sixth Amendment, which lowered the voting age from 21 to 18, on the political power of all young U.S. citizens and college students in particular.

How do we know an institution when we see one? In everyday language, physical places—for example, a jail—are sometimes referred to as institutions. But in sociology, *social institutions* are patterns of behavior governed by rules that are maintained through repetition, tradition, and legal support. Members of every society create and maintain social institutions to control human behavior and go about meeting their basic needs. How do we know what is a basic need for a society? Try creating your own imaginary society and thinking about what you need to do for it to survive. What means of controlling behavior would you establish to maintain your society? Keep in mind that you can eat what you like at home, but we all share the problem of where the garbage will go when you are done.

Imagine that you and your fellow students are stranded on a distant planet that looks and feels like Earth but has no other human inhabitants. What is the first thing you would do? Almost certainly, you would (a) figure out what you need to do to survive and (b) start assigning people to those tasks you determine need doing. The first thing you would probably need to do is ensure some semblance of order for these undertakings. So the first institution you would set up is some sort of *government*. (We're assuming that there are a lot of other students with you, not just a few classmates.) Second, you would have to start producing some food, finding water, and arranging for some system to distribute these goods. Whether you all share equally or distribute the goods according to some complex system of entitlements, you would be creating a system that includes ownership and "exchange value." In other words, you would be creating an *economic* institution, an institution that organizes how a society generates, allocates, and uses products and services. Because the needs of your group in this imaginary world would be different from the needs of the society in which you currently live, you would not need the same institutions in the same way. But here is the interesting part about ideas and practices that become *institutionalized*: It is hard to imagine them differently. For that reason, you would have a very hard time creating a government or an economy that did not strongly resemble the one you know now (or at least one that you have read or heard about). And yet, when you stop to think about it, there are innumerable ways in which societies *could* choose to organize their governments and economies.

Although not everyone believes in a higher power or practices a religion, every known society has had some form of religious institution (just ask an anthropologist). Would you and your companions adopt a unified system of belief to help you make sense of your new situation? If you were not rescued quickly, you might start trying to establish, in an organized way, a connection between yourselves and a higher power. Organizing a new religion or reestablishing an old one (from Earth) might help you come to terms with your situation, feel that someone (or something) was watching over you, and believe that eventually (even if only when you die) you would be going to a better place, seeing old loved ones, and so forth. If you remained in the distant world for more than a very short period of time, you would also have to set up rules about who could have sex with whom (to avoid nasty fights and to protect physically vulnerable members of the population) and (eventually) who should take care of the offspring of such unions (and how). In doing so, you would be establishing the institution of the *family*.

Finally, if you remain stuck on that planet, you would have to ensure that new members born into your society could learn your culture and the skills necessary to help your society survive. You would have to establish a social institution responsible for educating the members of your society. Once you have done so, you would have established the fifth basic institution found in almost every society, an *educational system*.

As you have probably already noticed, these institutions are all related to one another. A functionalist would maintain that they are also all *interdependent*. Just as a living organism suffers if a single one of its major organs (e.g., the kidneys or lungs or heart) starts to fail, functionalists maintain that if one institution is not working properly in a society, all the others (and the aggregate society) will suffer as well. For example, if our education system is not carrying out its function properly, young adults will not be prepared to get good jobs and therefore will not be able to support a family, pay taxes, financially support their religious organizations, or buy goods. Eventually, the faulty educational system would harm the family, economic, religious, and political institutions.

Unlike functionalists, conflict theorists examine the manner in which different interests in society work against one another. Karl Marx, for example, famously demonstrated that the worker class and the owner class were necessarily in conflict over just about everything that went on in society, from the organization of work to the proper use of the police and courts to the workings of a free press. But conflict also occurs among social institutions. For example, some religious institutions believe that they should have control over governmental and educational institutions, and throughout the history of the United States, there has been a seesaw power relationship between economic and governmental institutions.

Marx maintained that there was one institution that influences and largely directs all the other institutions. According to Marx (1859/1970), as the type of economic system changes, so does the makeup of the other institutions. The government, the schools, the family, and religion are all tools for those who own the means of production. Marxism, therefore, begins with the assumption of *economic determinism*—that is, that the economic institutions shape the rest. Although all the major institutions are tied to one another in some way, in this chapter, we will focus on the social institutions of the family and the economy, and we'll then focus on education, government, and religion in the following chapter.

The Family

Marx (1859/1970) maintained that relations among family members and even the average size of families are influenced by changes in the economic system. For example, under an agrarian economic system, in which work is centered on the family, the size of families is large so that they can produce many workers to till the land and produce crops. As societies become industrialized, and the work shifts to factories and other centralized locations, families become smaller. The move from farm to city means that families can no longer feed themselves through producing more food in the fields. More children would mean more mouths have to be fed through low-wage work in factories.

Today, this plays out in similar ways as during Marx's time. One notable change is that, while some of the work today resembles the manufacturing-based work of the past (working in factories), many low-wage workers today are employed in service sector work. In some ways, this makes them even more vulnerable as laborers, since manufacturing work, as dehumanizing and robotic as it might be on a daily basis, tends to be skilled labor, whereas service sector work often is less skilled. Service sector workers require less training and are more easily replaced than manufacturing workers, thus leaving them with less leverage to advocate for their rights and benefits. Today, 58.5 percent of all American workers are wage earners (meaning they receive hourly pay rather than an annual salary), and of these, 2.6 million workers earn minimum wage or less, and less than 1 percent of those working full-time in minimum wage jobs can afford to rent a one-bedroom apartment (BLS, 2016a; Bloom, 2017). Of all workers in the United States, 11.6 percent (1 out of every 8.5 workers) are paid poverty wages (Cooper, 2018). Wage issues are exacerbated for women, who earn less than men, and people of color, who earn less than whites, with the median wage for all men across professions in 2017 being $19.97 and for women $16.79, while the median wages for whites $20.10, blacks $14.99, and Hispanics $14.94 also indicate a significant gap (EPI, 2018).

According to Marx's writing partner, Friedrich Engels (1884/1942), families maintain the economic system of capitalism and the existing class structure. Legal marriages were created so that men would be able to know clearly who their heirs were so that they could bequeath their wealth to their male offspring. A marriage contract transferred a kind of ownership of the bride from her father, whose name she bore, to the husband, whose name she adopted. The traditional patriarchal family structure also worked to perpetuate the capitalist economic system because it allowed men to devote themselves to making money while their wives took care of them, their children, and their home (for no pay). This system also contributed to the maintenance of gender inequality.

Whereas conflict theorists use a macro-level approach to see connections between families and economic systems, symbolic interactionists use a micro-level analysis to focus on how institutions influence the roles men and women play and the status they assume in the family. As you learned in Chapter 9, even today and even in the most egalitarian nations in the world, gender socialization within families contributes to inequality between men and women.

In the United States, as throughout the world, women still do more of the household labor than do men. Even though many husbands and wives are likely to say that they should share their domestic chores equally, women still end up doing more of the work around the house. Daily, women spend 37 minutes on food and drink preparation while men spend 17, women spend 17 minutes doing laundry to men's 5 minutes, and while men spend 10 minutes daily on interior cleaning, women spend 29 minutes. Even when factoring in yard work and reparations around the house, on average men spend an hour and twenty-five minutes on household activities daily while women spend an average of two hours and 15 minutes, almost 6 hours more per week or 300 hours more per year (BLS, 2016b)!

There are many changes that the U.S. family is experiencing. For instance, the percentage of babies born to unwed parents has increased since the late 1990s. In 2012, 40 percent of all babies in the United States were born to unmarried women (Griffin, 2018). The pattern is clear: Fewer children are being raised by married parents. Many people assume the trend to be attributed to more young, single women having babies, but this is not the case. In fact, these data are more reflective of larger changes in U.S. society, such as LGBTQ couples, unmarried but relationship-secure heterosexual couples, and others having babies together. In other words, while the mother of many of these babies may be unwed, the children will be raised by two parents in new versions of the nuclear family unit.

Other changes to the institution of the family are now taking place. As same-sex couples and working mothers are becoming increasingly common, they are influencing the shape of families in the United States and across the globe. For example, as noted in Chapter 9, a growing number of U.S. states and nations across the globe recognize marriages or civil unions between same-sex adults.[2] Also, most mothers now work outside the home (as well as in it). Today, 73.2 percent of U.S. mothers with children under 18 years work outside the home. The percentage of employed mothers with children under the age of 6 has also increased dramatically over the past few decades (from 39 percent in 1975 to 61.5 percent in 2017) (Bureau of Labor Statistics, Employment Characteristics, 2018; NWLC, 2017). Even children raised by two married parents tend to have working mothers. In 2016, among married couples with children, 61 percent had both parents in the paid labor force (BLS, Employment in Families, 2017).

It is important to note that working both outside and inside the home is not a new phenomenon for women, particularly for women of color (Coontz, 1992). However, the fact that so many women of all races and social classes are now doing so is a relatively new social phenomenon. Today, the options families have, such as keeping one potential breadwinner at home, hiring a caregiver for their children, sending their children to day care facilities, and so forth, all relate to the economic position of the family. With both day care fees and household expenses, such as medical insurance, rising rapidly, parents with limited education and job skills face almost impossible choices when trying to both support their families financially and supervise and care for their children.

Family life can be even more expensive and stressful for parents who are part of the "sandwich generation," meaning they have both dependent children and elderly parents who need care. While grand-parents can be of great help to their adult children, they can also need assistance from them. As life expectancies rise and the baby boomers age, their adult children are often faced with figuring out how to assist their parents while raising their own children. Limits on prescription and health care coverage can leave many elderly people unable to pay for their medical needs or afford nursing assistance when they need care. Adult children, often with children of their own, must frequently face tough family and economic choices. The situation is even bleaker if they are uninsured. If that is the case, it is likely that they cannot even afford to take care of themselves! Nearly one third of all Americans with elderly parents provide financial help, and interestingly, families below the median income are more likely to provide financial support. While at first look, this may seem surprising, of course with generational inequal-ity families above the median income are more likely to come from fam-ilies above the median income who, in turn, are more likely to have been able to properly save for retirement (Pew Research Center, 2015). For those who do provide support for aging parents, of the $266,000 in average costs, 55 percent is paid out of pocket (Gleckman, 2017). Fortunately, the numbers of the uninsured saw a strong decline with the rollout of the Affordable Care Act (ACA), with 44 million uninsured in 2014 down to 27 million in 2016. Unfortunately, with changes of rules and decreased funding to create awareness about enrollment in the ACA under the Trump administration, the number of uninsured increased by 700,000 in just 2017 (Kaiser Family Foundation, 2018).

The Economic System

Economies organize how a society creates, distributes, and uses its goods and services. Today, we all live in a global economy in which the economic system of capitalism dominates. As Max Weber described it, the ratio-nal capitalism that arose in the West and that has spawned globalization requires paid (not slave) labor, free movement of goods, and sound legal and financial institutions. It also requires a bureaucratic system that hinges on educated administrators, accurate record keeping, and the technology to allow for long-distance communication and transportation (Collins, 1980). Goods and services are created and sold, for profit, across national borders at an increasingly rapid pace.

Inequality among nations is related to what each contributes to and takes from the global economy. "Global north" nations (most postindus-trial nations, such as the United States, Western European nations, and Japan) primarily contribute service work in the knowledge economy, with high-skilled workers (Odeh, 2010). In contrast, "global South" nations (e.g., most Latin American, African, Middle Eastern, and Asian nations) tend to produce raw materials or provide cheap labor to produce the goods consumed in global North nations. While some global South nations, such as India and China, are becoming increasingly competitive (the two nations are home to over 58 percent of recent college graduates in STEM subjects [World Economic Forum, 2016]), most global South nations are far from equal competitors to nations in the global North. In addition, corporations based in global North nations often control the resources of global South nations, and this makes it difficult for the poorer nations to build the infrastructure needed to create stronger economies. The nations with strong educational, banking, legal, and military systems have benefited the most from the globalization process and have used their economic advantage to increase their power in determining global governance and the future of the global economy. Other sectors of the world have not benefited as much from globalization. The World Trade Organization (WTO), which oversees the terms of global trade; the International Monetary Fund, which manages global financial markets; and the World Bank, which provides loans for economic development, are primarily controlled by global North nations and influenced by the concerns of global corporations, further disadvantaging the world's poorer nations.

Today, 15 percent of the world's population lives on less than $2 a day, and 71 percent (6.2 billion people) live on between two and ten dollars a day (Kochhar, 2015). Contrary to the oft-spoken retort that "that's a lot of money 'over there,'" the result is that there are nearly 815 million hungry people in the world, 844 million people who lack access to basic drinking water, 2.3 billion people without basic sanitation, and 1.1 billion people who have no access to electricity (World Food Programme, 2017; World Health Organization & UNICEF, 2017; Daly & Walton, 2017).

Moreover, more than half of the world's wealth is held by just the top 1 percent of wealthy people and the bottom fifty percent of people (3.7 billion people) hold just 1 percent of the world's wealth (Credit Suisse Research Institute, 2017). Further, in 2017, 82 percent of all the new wealth in the world went to just the top 1 percent of people, and just the new wealth gained by billionaires (not including any of the wealth they already had) would have been enough to end global extreme poverty seven times over (Oxfam, 2018)! Growing levels of economic inequality can also be seen *within* global North nations. In the United States, inequalities have steadily increased over the past 20 years, as the middle and lower classes have lost income and the very wealthy have gained greater income. In 2016, the *median* household income (meaning that half of all households made more and half made less) of all Americans was $57,617. In contrast, the average CEO made 271 times as much as the average worker, which means the average CEO makes more in one and a quarter days of work than the average worker will make for the year (Donnelly, 2017).

One of the most drastic outcomes of these trends in wealth and poverty is that 15 million households with 40 million people struggled to afford food in 2017 (they were *food insecure*), and more than 16 percent of all households with children in the United States were food insecure (Keith-Jennings, 2018). Many of the people who are hungry in the United States are members of the working poor. Workers on the low end of the global economic spectrum toil for very little compensation. The result is often hunger and deprivation for the workers and their families. Some work for even less than the minimum wage. Although sweatshops are most common in global South nations because of lack of unions and inadequately enforced (or nonexistent) labor laws, there are also pockets of sweatshops in global North nations, including many right here in the United States. The study described in Bernhardt et al.'s (2009) *Broken Laws, Unprotected Workers: Violations of Employment and Labor Laws in America's Cities* revealed that "many employment and labor laws are regularly and systematically violated" (p. 2) in the United States. The authors of this report surveyed workers in low-wage industries in Chicago, New York, and Los Angeles. They discovered that more than one out of four low-wage workers were not paid even the minimum wage during the previous week and more than three out of four of those who worked longer than 40 hours the week prior to the survey were not paid overtime wages as required.

Organized people have pressured their government to take action to address wage theft and other labor violations, such as those described above. As more and more organized citizens demand that their elected leaders do something to curb the abuse of workers, anti-sweatshop legislation has become increasingly common. Seven states, 45 cities, 16 counties, and 118 school districts have adopted sweat-free policies (see https://laborrights.org/our-work/sfc).

College students have been some of the most powerful opponents of sweatshops. On campuses across the nation, undergraduates have been striving to make sure that the apparel sold on their campus is made in sweat-free factories. As noted in Chapter 1 (this book), 193 colleges and universities (such as The College of the Holy Cross, Bridgewater State University, Beloit College, Indiana University, California State University at San Bernardino, and Miami University) have joined the Worker Rights Consortium (WRC) "to combat sweatshops and protect the rights of workers who make apparel and other products" (WRC, 2018).[3]

Sociologist in Action

Alan Ashbaugh

In the following paragraphs, Alan Ashbaugh, a recent college graduate, describes how he uses the sociological tools he gained as a sociology major at Colby College in Maine in working to bring about social justice locally, nationally, and globally.

Studying sociology greatly influenced the course of my life, giving me the tools to examine the social world, the skills and inspiration to effect positive change, and the direction to use my sociological training meaningfully and effectively.

In my sociology classes at Colby College, Waterville, Maine, I learned about the nature of society; about the harsh and complex realities of social issues; and, most important, that change toward a more just world is possible. A sociological truth that I will always remember is that society is not a static, unstoppable force, as it is so often referred to in popular media, but it is instead a constantly emerging and shifting combination of individuals making decisions and taking action. Thus, the world is already changing; we need only to guide it in the right direction by taking effective social action.

Of the many issues we explored in my introductory sociology course, the one that stood out for me is the vast economic inequality in my own community and internationally, as well as the social classes that create and perpetuate these inequalities. A subsequent sociology course on globalization broadened my sociological perspective to focus on the global issues of poverty, inequality, and justice, and the impact of each individual's actions on people worldwide. Examining these issues led me to a more vivid understanding of the social world; a stronger grasp of issues in my local and international communities; and a better sense of my place in Colby, Waterville, and society as a whole. This newfound understanding inspired me to get involved on and off campus, using my sociological training to move toward change. I became a founding leader of the Colby South End Coalition, which addresses the social and economic divide between Colby and its hometown of Waterville by encouraging volunteerism, dialogue, and a coming together of the two communities. I also took on the role as head of Colby Habitat for Humanity, which tackled economic inequality locally by connecting low-income families with decent, affordable housing.

(Continued)

(Continued)

Moving beyond Waterville in scope, I helped to organize the Voter Coalition to "get out the vote" for the upcoming presidential election, and I traveled to Chile to investigate the social promise of the emerging concept of corporate social responsibility.

When I graduated from Colby, I did not have a set career path. I only knew that I loved raising awareness of social issues and bringing students together to take action, and that I wanted to continue to be socially active beyond college. My senior year, I had the good fortune of learning about a job opportunity at an organization called Free The Children (now WE Charity) through an inspiring, socially involved professor in my sociology department. WE Charity was founded in 1995 by a 12-year-old boy named Craig Kielburger, who was so appalled at the practice of child labor he had read about in the newspaper that he gathered a group of his friends and classmates to raise awareness and take action to effect change—exactly what I was passionate about. In the past 10 years, WE Charity has grown to become the largest network of children helping children through education, having built over 400 schools, educating more than 35,000 children every day, having sent more than $9 million worth of medical supplies, and having implemented alternative income projects benefiting more than 20,000 people in developing countries. WE Charity's mission and the work they do matched my own personal goals and beliefs, so I applied for the position just prior to graduation and was hired! Today, as an International Youth Coordinator at WE Charity, I provide youth with the tools to make a difference in the world, such as information on social issues, effective fundraising and awareness-raising techniques, and the public speaking skills to powerfully communicate their passion to their peers. I use my sociological training each day in my work at WE Charity in examining the social issues that we are working to address, discovering the best ways to encourage and enable young people to effect positive change and "become agents of change for their peers around the world." The image of society I formed from my sociology classes of a constantly emerging collection of individual actions, in conjunction with the social issues we studied, help me to remember every day the power that each individual has to make a huge difference and the tremendous importance of doing so.

Source: Courtesy of Alan Ashbaugh.

Exercise 10.1
Functions of the Economy

Imagine that you live in a society in which the economic institution is not working properly (e.g., there is high unemployment or high inflation). According to the functionalist theoretical perspective, how might the decline in the economic institution affect the other institutions in society? Specifically, how would it affect the (a) family, (b) educational, (c) religious, and (d) governmental institutions?

Now imagine that you live in a society in which the economic institution is strong and thriving. According to the functionalist theoretical perspective, how might the thriving economic institution affect the other institutions in society? Specifically, how would it affect the (a) family, (b) educational, (c) religious, and (d) governmental institutions?

Exercise 10.2
The Changing U.S. Family

There are multiple structural and cultural causes for the changes in the institution of the family that are noted above. There is also ongoing debate as to whether these changes are more positive or negative for society. Write a one- to two-page essay that examines these changes in the family from either (a) a conflict perspective or (b) a functionalist perspective. (If need be, refer back to the descriptions of these perspectives in Chapter 2.)

When writing your essay, be sure to

1. provide an overview of how society operates according to the theoretical perspective you have chosen and then apply that perspective to the changes in the social institution of the family, and

2. discuss whether someone coming from the theoretical perspective you have chosen would approve or disapprove of the changes in the family (and why or why not).

Exercise 10.3
Making a Living Wage

A living wage differs from a minimum wage in that it accounts for the minimum income needed for a worker to provide basic needs such as food, housing, clothing, healthcare, and other essentials. Often times, this more realistic way of calculating the wage needed by workers is higher than minimum wage.

1. Read "Section One: What Is the Living Wage?" in "Living Wage Policy: The Basics" (https://www.epionline.org/wp-content/studies/epi_livingwage_08-2000.pdf).

2. Visit the website for the Living Wage Calculator at http://livingwage.mit.edu/ and calculate the living wage necessary for a family of four living in the county where your university is based.

3. Visit the website for the Family Budget Calculator at https://www.epi.org/resources/budget/ and determine the family budget necessary for a family of four living in the county where your university is based.

4. Write a two-page paper outlining what you've learned from these two websites, analyzing it through the lens of the research you've conducted for this exercise on the living wage. Make sure to also add your own sociological analysis and any policy suggestions you want to put forward.

Exercise 10.4
The World Trade Organization

1. Go to the Public Citizen website and read about the World Trade Organization (WTO) (https://www.citizen.org/sites/default/files/wtochapter1_0.pdf) and then go to the WTO's own website and read "What Is the WTO?" (http://www.wto.org/english/thewto_e/whatis_e/whatis_e.htm).

2. Read the descriptions of the WTO on both websites.

3. Write a paper that answers either Questions a to c or Questions d and e below.

   a. What groups or organizations are represented in each document?

   b. As a sociologist, which documents do you think are most convincing? Explain why.

   c. Did you know very much about the WTO before reading this chapter and these documents? If yes, what gave you that knowledge? If not, why do you think you didn't know much about the WTO?

   d. How would a conflict theorist describe the function of the WTO?

   e. How would a functionalist theorist describe the function of the WTO?

   Extra credit: Answer all five of the above questions (a–e). Then, outline what you and your classmates could do to create better awareness on campus and in your local community about the WTO in an effort to help put pressure on it to operate in a more transparent and equitable manner.

Exercise 10.5
When It's American Workers Who Are Exploited

Read "Ikea's U.S. Factory Churns Out Unhappy Workers" at http://articles.latimes.com/2011/apr/10/business/la-fi-ikea-union-20110410 and write a two- to three-page paper in which you answer the following questions:

1. How does the information in the article relate to the material in this chapter?

2. How does it make you feel to read about how a Swedish corporation treats workers in the United States far worse than its workers in Sweden?

(Continued)

(Continued)

3. Do you think that these inequities should be addressed and, if so, by whom? If not, why not?

4. Would you hold U.S. corporations to the same standards when they employ workers in foreign countries as you are suggesting we hold Ikea to when employing U.S. laborers? Why or why not?

5. How did this article and the chapter influence your perspective on global inequality? Why?

Exercise 10.6
Why Are We Cutting Food Stamps?

Read and watch the related video "Cut in Food Stamps Forces Hard Choices on Poor" at http://www.nytimes.com/2013/11/08/us/cut-in-food-stamps-forces-hard-choices-on-poor.html?pagewanted=2&_r=0&src=rechp and read "Food Stamp Cut Backed by Republicans With Voters on Rolls" at https://www.bloomberg.com/news/articles/2013-08-14/food-stamp-cut-backed-by-republicans-with-voters-on-rolls and "The Insanity of Our Food Policy" at http://opinionator.blogs.nytimes.com/2013/11/16/the-insanity-of-our-food-policy/, then answer the following questions. Can you relate to the lives of the people described in these articles and the video? Why or why not?

1. Were you aware that one in seven Americans relies on food stamps? What is your reaction to that fact? Why?

2. How does cutting the federal budget for food stamps negatively affect the economy, as described in the articles and the video? Who benefits from such cuts?

3. What theoretical perspective described in this book best explains why some poor people, who rely on food stamps, vote for members of Congress who support cutting food stamp benefits? Why?

4. The Trump administration has consistently proposed cuts to the Food Stamps program and has worked to upend some of the provisions of the Farm Bill geared toward alleviating hunger. Read "Trump Administration Works to Restrict Food Stamp Access the Farm Bill Protected" at https://www.nytimes.com/2018/12/20/us/politics/food-stamps-trump-administration-snap.html and offer your thoughts and analysis.

5. Describe what steps you can take to curb hunger, specifically through policy and structural level solutions, in (a) your local community, (b) your state, and (c) the nation.

Exercise 10.7
Students Creating Change: The Anti-Sweatshop Movement

Go to the United Students Against Sweatshops website (http://usas.org).

Write a two-page paper that describes (a) one campaign on which United Students Against Sweatshops is currently working, (b) whether or not you could see yourself joining that campaign, and (c) why you could or could not see yourself participating in the campaign.

Exercise 10.8
Young People in the Lead

1. Go to the website for the Millennium Campus Network (MCN) (https://www.mcnpartners.org/) and familiarize yourself with the work of the organization as well as the work of Sam Vaghar, the young founder and executive director of MCN (https://www.mcnpartners.org/sam-vaghar/).

(Continued)

(Continued)

2. Similarly, go to the websites for Generation Citizen at https://generationcitizen.org/, the Ben and Jerry's Foundation at https://benandjerrysfoundation.org/the-grassroots-organizing-for-social-change-program/, and Oxfam On Campus at https://www.oxfamamerica.org/take-action/volunteer/campus/ and familiarize yourself with the work of these organizations.

3. What did you learn when reading through these websites? What programs, initiatives, or information excited you? Did you see anything that you want to join or apply for? What did you learn about student action and activism from these organizations that either changed or increased your thinking about students organizing for social change?

Exercise 10.9
Immigration and Citizenship

1. Select two nations outside the United States (one in the Western Hemisphere and one in the Eastern Hemisphere), and go to the U.S. Citizenship and Immigration Services home page (http://www.uscis.gov/portal/site/uscis).

2. Write a two-page paper, using the information on the website, to figure out how you might successfully emigrate from each respective nation to the United States. Be sure to include the factors that would ease your path to U.S. citizenship (being a highly skilled worker, being wealthy, having a family member here, etc.).

3. Conclude your paper by describing what you learned from the exercise and how it affected your perspective on the immigration process. (If you did not learn anything new, explain why you did not.)

4. Read "Immigration Under Trump: A Review of Policy Shifts in the Year Since the Election" (https://www.migrationpolicy.org/research/

immigration-under-trump-review-policy-shifts) and "90-Year-old Ethel Kennedy Joins Hunger Strike Against Immigration Policies" (https://www.bostonglobe.com/metro/2018/06/26/year-old-ethel-kennedy-joins-family-members-hunger-strike-against-immigration-policies/rJY29ksRc9UZrww0n2hv0N/story.html).

5. Write a two-page paper describing your reflections and analysis after having read each of the above.

6. Extra Credit: Read "The Right to Immigration Institute Offers Workshops, Legal Services" (https://www.bostonglobe.com/metro/regionals/west/2018/05/02/the-right-immigration-institute-offers-workshops-legal-services/CPw6dpYoQwDtJpqmZb6RfM/story.html) and look through the organization's website (https://www.therighttoimmigration.org/). This organization was started up and is run (as of 2019) by college students. Add one to two pages to your paper (above). What does learning about Right to Immigration and the students who started and run it teach you? What about their work do you find interesting, and what might you change or add to it if you were in charge? Does reading about their work make you consider that you also have the potential to start a program or organization that works to create social change?

DISCUSSION QUESTIONS

1. What role do you think economic globalization plays in legal and illegal immigration to the United States? Push yourself to think on structural/societal levels as you answer.

2. How do you think the changes in our economic institution (e.g., inflation, fewer jobs that pay a "family" wage, the growing need for a college education) are affecting the social institution of the family?

3. Why is it in the interest of global North workers to organize with global South workers for higher wages?

4. What families might continue to be most vulnerable to the negative impacts of the Great Recession? Why? Do you feel that your family has

been affected by the recession? If not, why not? If yes, why and in what ways?

5. Which theoretical perspective, conflict theory or functionalism, would you choose to best make sense of how the economic institution operates in U.S. society? Why?

6. Imagine that the institution of the family suddenly disappeared. How might U.S. society be different without this primary institution? Do you think U.S. society could survive without it? Why or why not?

7. If you marry or begin a domestic partnership, do you think you will share domestic chores equally with your spouse? Why or why not? Be sure to discuss how your gender socialization, your family of origin, and your place in the economic institution affect your answer. Or if you are already married or have begun a domestic partnership, how do these factors affect whether or not you share domestic chores equally?

SUGGESTIONS FOR SPECIFIC ACTIONS

1. Research the family-related personnel policies at your college. Do you think they provide adequate benefits for employees who must care for a sick family member? If not, develop a reasonable suggestion for policy reform, and work with your student government leaders to present your ideas (and advocate for them) to the administration.

2. Research the policies that your college has for providing benefits to partners of lesbian, gay, bisexual, transgender, or intersexed (LGBTI) employees. Are they the same as for the partners of heterosexual employees? If not, why not? How do your findings, in either case, reflect the values of your college (*your* institution!) and of the greater society? If you find inequalities and believe that they need to be changed, develop a reasonable suggestion for policy reform, and work with your student government leaders to present your ideas (and advocate for them) to the administration.

3a. Review the information on the U.S. Immigration and Customs Enforcement's Students and Exchange Visitors webpage (http://www.ice.gov/sevis/students).

3b. Interview the administrator on your campus who is in charge of international students. Ask him or her how your school assists international students through the process of becoming students on your campus.

3c. Interview five international students at your school. Ask them about the steps they went through to become students there. Be sure to note any differences in their experiences and why these differences may have occurred (the year of their matriculation, the nation from which they came, their economic status, etc.).

3d. Use this information to create a resource for international students that could be put on your school website (if approved by your college or university administration).

4. Research the salaries and benefits of the cafeteria workers and custodial professionals on your campus. Do they receive a living wage? Do some research to learn more about how students on other campuses have advocated for cafeteria workers and custodial professionals and how you might do so as well. The link http://www.jwj.org/ is a starting place to find information.

5. Go to WE Charity's website (https://www.we.org/). After browsing through it, read the information provided on child labor (https://www.we.org/we-schools/program/issues-backgrounders/global-child-labour/). One of the most powerful ways to fight against child labor is to ensure that children have schools to attend and the resources (funds for school fees, books, uniforms, etc.) to afford to go to school. Now that you know about child labor and how important education is, you can start a campaign to raise money to build schools in regions known for child labor. Check out the resources at https://www.we.org/we-villages/education/ and begin a school-building campaign at your college.

Please go to this book's website at http://study.sagepub.com/white6e to find further civic engagement opportunities, resources, peer-reviewed articles, and updated web links related to this chapter.

NOTES

1. Of course, some students leave high school and join the military institution. Many do this in the hope of receiving the training and experience they need to succeed economically once they leave the military.
2. The Internal Revenue Service now recognizes same-sex marriages.
3. For more information about the campaign, go to the WRC website (http://workersrights.org).

REFERENCES

Bernhardt, A., Milkman, R., Theodore, N., Heckathorn, D., Auer, M., DeFilippis, J, & Spiller, M. (2009). *Broken laws, unprotected workers: Violations of employment and labor laws in America's cities*. Retrieved from http://www.nelp.org/page/-/brokenlaws/BrokenLawsReport2009.pdf?nocdn=1

Bloom, E. (2017, July 14). Only 0.1% of US minimum-wage workers can afford a 1-bedroom apartment, report finds. Retrieved from https://www.cnbc.com/2017/07/14/only-point-1-percent-of-us-minimum-wage-workers-can-afford-a-1-bedroom.html

Bureau of Labor Statistics. (2016a, April 01). Characteristics of minimum wage workers, 2015. Retrieved from https://www.bls.gov/opub/reports/minimum-wage/2015/home.htm

Bureau of Labor Statistics. (2016b, December 20). American time use survey. Retrieved from https://www.bls.gov/tus/charts/household.htm

Bureau of Labor Statistics. (2017, April 27). Employment in families with children in 2016. Retrieved from https://www.bls.gov/opub/ted/2017/employment-in-families-with-children-in-2016.htm

Bureau of Labor Statistics. (2018, April 18). Employment characteristics of families – 2017. Retrieved from https://www.bls.gov/news.release/pdf/famee.pdf

Collins, R. (1980). Weber's last theory of capitalism: A systematization. *American Sociological Review, 45*, 925–942.

Coontz, S. (1992). *The way we never were: American families and the nostalgia trap*. New York, NY: Basic Books.

Cooper, D. (2018, June 15). One in nine U.S. workers are paid wages that can leave them in poverty, even when working full time. Retrieved from https://www.epi.org/publication/one-in-nine-u-s-workers-are-paid-wages-that-can-leave-them-in-poverty-even-when-working-full-time/

Credit Suisse Research Institute. (2017). Global wealth report 2017. Retrieved from http://publications.credit-suisse.com/tasks/render/file/index.cfm?fileid=168E2808-9ED4-5A5E-19E43EA2A731A4ED

Daly, H., & Walton, M. A. (2017). Energy access outlook 2017: From poverty to prosperity. *International Energy Agency*.

Donnelly, G. (2017, July 20). Top CEOs make more in two days than an average employee does in one year. Retrieved from http://fortune.com/2017/07/20/ceo-pay-ratio-2016/

Economic Policy Institute. (2018). *State of Working America Data Library* [Data file and code book]. Available from Economic Policy Institute Web site at https://www.epi.org/data

Engels, F. (1942). *The origin of the family, private property, and the state* (F. L. Morgan, Trans.). New York, NY: International. (Original work published 1884)

Gleckman, H. (2017, January 18). Families spend more to care for their aging parents than to raise their kids. Retrieved from https://www.forbes.com/sites/howardgleckman/2017/01/18/families-spend-more-to-care-for-their-aging-parents-than-to-raise-their-kids

Griffin, R. (2018, October 17). Almost half of U.S. births happen outside marriage, signaling cultural shift. Retrieved from https://www.bloomberg.com/news/articles/2018-10-17/almost-half-of-u-s-births-happen-outside-marriage-signaling-cultural-shift

Kaiser Family Foundation. (2018). Key facts about the uninsured population. Retrieved from https://www.kff.org/uninsured/fact-sheet/key-facts-about-the-uninsured-population/

Keith-Jennings, B. (2018, September 5). Millions still struggling to afford food. Retrieved from https://www.cbpp.org/blog/millions-still-struggling-to-afford-food

Kochhar, R. (2015). Seven-in-ten people globally live on $10 or less per day. Retrieved from http://www.pewresearch.org/fact-tank/2015/09/23/seven-in-ten-people-globally-live-on-10-or-less-per-day/

Marx, K. (1970). *A contribution to the critique of political economy* (M. Dobbs, Trans.). New York, NY: International. (Original work published 1859)

National Women's Law Center. (2017, April). Child care: A snapshot of working mothers. Retrieved from https://nwlc.org/wp-content/uploads/2017/04/A-Snapshot-of-Working-Mothers.pdf

Odeh, L. E. (2010). A comparative analysis of global North and global South economies. *Journal of Sustainable Development in Africa*, 12(3), 338–348. Retrieved from http://www.jsd-africa.com/Jsda/V12No3_Summer2010_A/PDF/A%20Comparative%20Analysis%20of%20Global%20North%20and%20Global%20South%20Economies%20(Odeh).pdf

Oxfam International. (2018, January 22). Richest 1 percent bagged 82 percent of wealth created last year—poorest half of humanity got nothing. Retrieved from https://www.oxfam.org/en/pressroom/pressreleases/2018-01-22/richest-1-percent-bagged-82-percent-wealth-created-last-year

Pew Research Center. (2015, May 21). *Family support in older societies*. Retrieved from http://www.pewsocialtrends.org/2015/05/21/family-support-in-graying-societies/

Worker Rights Consortium. (2018). *About us*. Retrieved from http://www.workersrights.org/about/as.asp

World Economic Forum. (2016). *Human capital report 2016*. Retrieved from http://reports.weforum.org/human-capital-report-2016/learning-through-the-life-course/

World Food Programme. (2017, September 15). *World hunger again on the rise, driven by conflict and climate change, new UN report says*. Retrieved from https://www.wfp.org/news/news-release/world-hunger-again-rise-driven-conflict-and-climate-change-new-un-report-says

World Health Organization, & UNICEF. (2017). *Progress on drinking water, sanitation and hygiene: 2017 update and SDG baselines*.

Social Institutions, Continued

Education, Government, and Religion

S hould students in U.S. public schools be required to "pledge allegiance to the flag of the United States of America and to the Republic for which it stands, one nation under God, indivisible, with liberty and justice for all"?[1] This question reveals the connection among educational, political, and religious institutions in the United States. How is our nation connected to God? Which God? Whose God? Are you less of an American if you don't believe in God (or if you do)? Why did Thomas Jefferson and the other authors include the concept of "building a wall of separation between Church & State" as they developed the First Amendment to the United States Constitution (Jefferson, 1802)? These questions have been debated by politicians, religious leaders, and educators across the United States. In this chapter, we examine the institutions of education, government, and religion and their relationships to one another and to the other primary institutions in U.S. society.

Educational Institutions

Educational institutions teach young members of society the basic and (in some cases) advanced skills needed to function effectively in our society. In turn, educated citizens enable societies to run smoothly and to become more technologically advanced, more productive, and more prosperous. The educational level of a nation's population is directly related to its level of income equality and overall economic health.[2] As other key institutions in society change, the functions of the educational system must adjust accordingly. For example, as economies change, so do the skills needed by citizens and those taught in the educational systems. As institutions become increasingly complex and bureaucratized, they need an increasingly well-educated workforce. As Max Weber (1970) pointed out, "Trained expertness is increasingly indispensable for modern bureaucracies" (p. 240).

A 2016 Organisation for Economic Co-operation and Development (OECD) study shows, however, that the United States has fallen behind most other developed nations in teaching its population the technical and mathematical skills needed in the workplace today and is only average in literacy. The tests, in literacy, basic math, and "problem solving in technology-rich environments," were given to people of ages 16 to 65 in

40 OECD nations, Cyprus, and the Russian Federation. Young Americans scored lower than older Americans on the tests, and test takers in Japan and Finland earned the highest scores on all three tests (OECD, 2018).

There were also more high *and* more low scores among U.S. test takers than among those from other nations. So it appears that the U.S. education system is polarized, with a large divide between those at the top and those at the bottom, and with those in the middle learning fewer skills than students from most other developed nations. This does not bode well for the economic or social health of the United States or its individual residents. As the OECD (2013) report notes, "If there is one central message emerging from this new Survey of Adult Skills, it is that what people know and what they can do with what they know has a major impact on their life chances" (p. 6). Educational institutions must teach students information *and* how to use that information in the modern workplace, and they must do so for *all* citizens and not just those at the top of the educational ladder. The Nation's Report Card, a long-term project of the U.S. government, bears the bad news that the average U.S. student is less than 50 percent proficient in civics, geography, mathematics, reading, science, U.S. history, and writing, and this holds true for students tested in fourth, eighth, and twelfth grades (the study only examines these three grade levels) (National Assessment of Educational Progress, 2018). The news is unfortunately worse, when accounting for race, with white students gaining better quality education and better outcomes from an early age and continuing all the way through the education system.

Educational institutions are also connected in a variety of ways to the institution of government. Those who score higher on the OECD Survey of Adult Skills are also more likely to trust others, engage in volunteer activities, believe that they have an impact on political processes, and be employed and make above-average salaries (OECD, 2013). Schools teach students how government functions and their obligations as citizens to participate in public life. An educated citizenry also provides the skills and brainpower a nation needs in an increasingly competitive global society. To provide such a workforce, our educational system must be effective and properly funded.

State and local governments supply most of the funding for public schools. States provide 47 percent of the funding, local communities 45 percent, and the federal government 8 percent (National Center for Education Statistics, 2018). In most states, primary and secondary public schools receive funding based largely on local property taxes. Both state and federal governments create educational goals. State governments and education experts across the nation developed the Common Core Standards, which 41 states, 4 U.S. territories, and Washington, D.C., have now begun to implement. For the first time, almost every state will expect students to achieve the same standard of achievement in English and mathematics. In recent years, the federal government has also played a larger role in regulating the function and operation of public schools. For example, the federally mandated No Child Left Behind Act of 2001 (NCLB) established, for the first time, test-focused goals and timetables that public schools must adhere to in order to receive federal funding. The NCLB Act came under criticism for narrowing the curriculum, lowering standards to increase the numbers of passing scores, and focusing on test scores rather than overall educational growth (U.S. Department of Education, 2011). In response to this critique, the secretary of education in the Obama administration, Arne Duncan, granted states flexibility on NCLB requirements "in exchange for rigorous and comprehensive state-developed plans designed to improve educational outcomes for all students, close achievement gaps, increase equity, and improve the quality of instruction" (U.S. Department of Education, n.d., para. 1).

Despite the Common Core Standards and the NCLB Act, though, most of what occurs within our schools is still determined or administered by local government. School boards, whose members are locally elected or appointed, oversee the operation of schools and create goals at the local level that are consistent with state and federal laws. Even private schools, although they do have greater freedom in determining the curriculum, are bound to abide by U.S. laws, such as those that protect against discrimination. In short, government input can be found at all levels of the institution of education, in school curricula, school buildings, and teaching approaches.

All three of the institutions examined in this chapter—education, government, and religion—are evident when we look at the NCLB Act[3] and the school boards that are working to put this legislation into action. One of the most heated school board battles in recent years has centered on whether to teach "intelligent design" or evolution in science classes. Proponents of the faith-based idea of intelligent design—the notion that some intelligent, supernatural force was responsible for the creation of the world—have tried to influence the composition of school boards in their efforts to have this belief represented in the public school curriculum. The elected State Board of Education in Kansas, for example, changed that state's science curriculum to include the Christian notion of intelligent design in science classes alongside the teaching of evolution, whereas residents of Dover, Pennsylvania, voted to oust all eight of their local school board members after the board proposed to do the same thing there. But remember the First Amendment and the separation of church and state? This requires that schools teach based on the science and facts rather than religious (or other) ideologies and belief systems.[4]

In 2013, a physics professor at Ball State University who had been teaching intelligent design in his classes was prohibited from doing so. In making the decision, the president of the university, Jo Ann Gora, said,

Teaching intelligent design as a scientific theory is not a matter of academic freedom—it is an issue of academic integrity. The scientific community has overwhelmingly rejected intelligent design as a scientific theory. Therefore, it does not represent the best standards of the discipline as determined by the scholars of those disciplines. Said simply, to allow intelligent design to be presented to science students as a valid scientific theory would violate the academic integrity of the course as it would fail to accurately represent the consensus of science scholars. (Quote found in Kingkade, 2013)

Science rather than religion must determine what is taught in science classes at Ball State University.

Fundamentalist religious sources, however, have had a tremendous impact on what is taught in high schools. A 2011 poll of high school science teachers revealed that just 28 percent consistently follow the National Research Council's recommendations "to describe straightforwardly the evidence for evolution and explain the ways in which it is a unifying theme in all of biology" (Bakalar, 2011, para. 2), and 13 percent teach their students through the lens of creationism (e.g., intelligent design). The other 60 percent tend to teach both evolution and creationism and/or try to avoid the topic of evolution (which is at the heart of all biology!) in order to avoid controversy (Bakalar, 2011). According to a 2015 Pew Research Center poll, just 62 percent of Americans believe that "humans and other living things have evolved over time" (Pew, 2015). No doubt, this stems in large part from the fact that so few people learn the facts about evolution in their high school biology class.

Many young people have joined the effort to move religion out of science classrooms. For example, one high school student from Louisiana, Zack Kopplin, used his senior project to start a drive to repeal the 2008 Louisiana Science Education Act, which allows for the teaching of creationism in Louisiana public schools. Now a graduate of Rice University, Zack is widely known as a highly respected advocate for science. Zack's efforts have been endorsed by 78 Nobel laureates, the American Association for the Advancement of Science, and the New Orleans City Council (Dvorsky, 2013; Quenqua, 2013).

Government

The U.S. Constitution sets the framework for governance in the United States. As anyone who has ever taken a U.S. government or history course (or watched *Schoolhouse Rock*)[5] knows, the Preamble to the Constitution describes what the founders of this nation hoped to gain from establishing their government under the Constitution.

> We, the people of the United States, in order to form a more perfect Union, establish justice, insure domestic tranquility, provide for the common defense, promote the general welfare, and secure the blessings of liberty to ourselves and our posterity, do ordain and establish this Constitution for the United States of America. (See a transcript of the Constitution of the United States at http://www.archives.gov/exhibits/charters/constitution.html)

As the Preamble indicates, the government regulates a wide range of interactions and processes in our society. Government is responsible for the smooth functioning of society. This broad mandate includes ensuring public safety—from personal safety in neighborhoods, to infrastructure, such as making sure that levees are properly built and maintained to protect citizens from flooding,[6] to overseeing public commerce (everything from an efficient transportation system to stable financial markets), to running fair and democratic elections. The government, in short, is responsible for overseeing the well-being and social welfare of the nation's citizens.

While voting is the most important right of citizens living in a democracy, the majority of Americans do not actually exercise the power they have to elect their governmental representatives. Many do not even know who their elected representatives are, never mind whether or not they are actually "promot[ing] the general welfare." Just over half of all Americans who are eligible to vote exercise that right in presidential elections, with 55.4 percent voting in the 2016 presidential election and 50.3 percent in the 2018 midterm elections (Wallace, 2016; United States Elections Project, 2018). Even fewer people tend to vote in elections when only local government representatives or initiatives are on the ballot.

Many Americans feel disconnected from the political process and turned off by the huge amounts of money spent on the campaigns they see covered on television. However, unlike national or statewide elections, which involve millions of dollars and depend in great part on who can raise the most money, local elections are more open to people who commit the resources of time and energy to cultivating local contacts and to urging people personally and in small groups to vote for them. Therefore, whereas organized money is more important in state and federal elections, in local elections, organized people can propel someone into office. However, in all cases, those running for election or reelection to political posts can safely ignore people who do not involve themselves in politics and who are not likely to vote.

Religious Institutions

Approximately 75 percent of Americans identify as Christian. The term *Christian* encompasses a plethora of religions, ranging from mainstream Protestant to Evangelical Protestant to Catholic. Following the umbrella group of Christians, the second largest group related to religion is the unaffiliated. The percentage of unaffiliated has doubled in the past quarter century and continues to rise, with 36 percent of adults under 30 falling into this category (Newport, 2017; Pew Research Center, 2014).

Analyses of religious institutions differ rather dramatically among the three primary theoretical perspectives in sociology. These theories help us understand the different roles religions play in society, how religions adapt to changes in society, and how they influence society and the individuals in it. It is important to note that sociologists are interested in the *interaction between religion and society* rather than in the veracity of the teachings of a particular religion.

Following Durkheim's famous writings in this area, functionalists maintain that religion serves several functions for society: It unites its followers, helps establish order by providing a "correct" way of living, and gives people a sense of meaning and purpose in their lives. Of course, this particular functional analysis of religion assumes that there is only one religion in the society and does not consider societies with multiple religions, such as the United States. Religious organizations also often provide volunteer opportunities for members and socialize them into a lifestyle that includes volunteering and active involvement in their communities (Jansen, 2011; Johnston, 2013).

Conflict theorists argue that religions tend to distract oppressed people and prevent them from concentrating on the inequities of their societies. Marx once described religion as "the opiate of the people" (as quoted in Bottomore, 1964, p. 27). He argued that religion helps maintain the status quo by encouraging workers to ignore their sufferings here on Earth, thus acting as an "opiate" by keeping the oppressed subdued and uninterested in rising up against their oppressors. The clergy, financially supported by the owner class, counsel their followers not to protest the inequality that led to their poor conditions but rather to be docile and "good" here on Earth so that they may receive their reward for their earthly suffering and forbearance in Heaven.[7] Marx believed that "the abolition of religion as the illusory happiness of the people is the demand for their real happiness. To call on them to give up their illusions about their condition is to call on them to give up a condition that requires illusions" (Marx, 1844).[8] Marx contends that humans create religion in order to make sense of their human situation and in order to provide a higher "reason" for the unequal distribution of wealth, power, and resources in society. In his view, while religion is an attempt of humans to find happiness, real happiness can only be achieved if religion and other institutions aimed at explaining away the unfair circumstances of society are dissolved, and people focus instead on changing the structures that create inequality.

Symbolic interactionists point out that religions are socially constructed, created, and re-created by followers through the use of symbols and rituals. They agree with functionalists that religion gives order to our lives. However, they also stress that individuals have agency (the ability to create and change institutions) and play active and consistent roles in the design and maintenance of their religion. Just as societies change and people create, adjust, and discard symbols, religious institutions change as well. For example, the change in symbols and attitudes within the Catholic Church brought about by the Second Vatican Council (1962–1965) changed the outlook of the church and many of its rules. The language of the Mass changed from Latin to the tongue of the parishioners, the priest no longer faced away from his parishioners when saying Mass, and guitars and folk groups replaced many organs and choirs, respectively. These symbolic shifts changed how members of the church related to one another and saw their roles within the church. Just changing certain symbols helped foster a rise in status of the Catholic laity and created a Catholic Church that Catholics who lived 100 years ago would scarcely recognize.

The Relationships Among Educational, Governmental, and Religious Institutions

Like all other major institutions in society, educational, governmental, and religious institutions must find a way to coexist. Depending on one's theoretical perspective, the grounds of this coexistence can vary widely. For example, functionalists maintain that they can cooperate for the good of the whole society, whereas conflict theorists argue that the educational, governmental, religious, and family institutions merely carry out the desires of those who control the economic institution.

No matter the theoretical perspective, it is clear that social institutions must interact with one another in a variety of ways. However, *how* they interact with one another differs from one society to the next. For example, the influence of religion on government and educational systems varies widely from society to society. The laws of some societies are based on religious doctrines. In other societies, religions are not even officially acknowledged by the government.

The debate about religious garb in schools reveals how the relationships among government, religion, and education vary from society to society and over time. For instance, since 2004, France has prohibited all religious forms of dress, including hijabs (head scarves worn by Muslim girls and women), in schools. In 2011, Azerbaijan imposed a similar ruling, as increasing numbers of its citizens began to embrace Islam after the nation emerged from the forced secularism of the Soviet era (Abbasov, 2011). Meanwhile, in Saudi Arabia and Iran, women and girls *must* wear the hijab or face punishment from the government for not doing so (Talley, 2011), and Turkey has recently lifted its ban on the hijab in state institutions (Reuters, 2013). In each of these cases, government officials recognize schools as the socializing agents they are.

In states such as Saudi Arabia and Iran, where religious leaders have a strong influence over government, schools are used to promote the dominant religion. Likewise, the recent lifting of the hijab ban in public institutions in Turkey reflects the Islamist-based political party now in power in that nation. However, in nations with secular governments, such as France and Azerbaijan, that fear the rising influence of religion in society, schools are used to dampen the influence of religion. Meanwhile, in the United States, with our declaration of religious freedom and separation of church and state, we continue to debate whether creationism can or should be taught in public schools. As these examples illustrate, the social institutions of religion, education, and government—like all major social institutions—are interdependent and continually interact with and influence one another in every society.

Sociologist in Action

Tim Woods

Dr. Tim Woods grew up in a small farming community in Oklahoma. As an undergraduate majoring in sociology at Southwestern Oklahoma State University, he joined a group of two sociology professors and a handful of students who volunteered at a local prison. In the prison, the faculty and students served as associates and advisers to the Lifers' Club, an organization of prisoners who were serving life sentences. While the prison volunteer program had no formal ties to the sociology curriculum, this experience left a lasting impression on Woods's work and teaching.

Interacting with the inmates forced Woods to question some fundamental ideas he held, especially those concerning the roles of individual responsibility and self-efficacy. A key event concerned a very religious inmate Woods had befriended.

(Continued)

(Continued)

While many inmates joked that no one in the prison was guilty, this particular inmate vociferously and continually proclaimed his innocence, explaining that God had sent him to prison to help the other inmates. Wary of the inmate's stories, Woods understood these proclamations to be psychological dissonance, either a denial of individual responsibility or mental illness. However, some two years after he left the prison for graduate school, Woods received word that new evidence had been discovered establishing the inmate's innocence. He was released from prison. This became an important example to Woods of the real-world connections among social structure, power, and the lived experiences of individuals.

Dr. Woods is now the chair of the Department of Political Science and Sociology at Manchester Community College in Manchester, Connecticut. Wishing that his students have the same type of opportunity to learn and contribute to their community that he had had at the prison, Woods began embedding community engagement into the formal coursework in sociology. He wanted students to develop a sociological imagination through their interaction with the community. As they served in and reflected on life in their community, students began to connect first their own and then others' personal troubles to social issues in their community and the larger world. This experience also advanced students' understanding of broader sociological concepts and theories by visibly connecting those concepts to their own lived experiences and those of others in the community.

Initially, Woods's students simply volunteered at a variety of community nonprofit organizations and related their experiences there to the course material. Over time, Woods's and his students' roles grew from merely volunteering to publicly advocating for structural community change. Their sociological imagination enabled them to recognize the need for and potential of collective action to address the social needs of the community. Joining forces with the local homeless shelter, religious organizations, and town officials, they helped establish the Manchester Initiative for Supportive Housing and have taken a leading role in advocating for supportive housing as a long-term solution to homelessness. Their strategy for community awareness has become a topic of discussion among supportive housing advocates as far from Connecticut as California.

Woods believes that community colleges provide a unique opportunity for sociologists interested in teaching about and engaging in public sociology and community change. Unlike most students in four-year colleges and universities, community college students are more strongly embedded in their communities, where they live, work, worship, and go to college, and as a result, they have long-term stakes in the community. While it is certainly possible for professors and students at four-year schools to make a positive impact on the communities surrounding their campus, they must often make extra efforts to know and be accepted by members of the local community. Tim Woods's students, similar to Jane Addams and the other Hull House researchers and activists, live in the community they are trying to improve. Also, community college students are usually more representative of the diversity (in terms of race and ethnicity, class, age, religion, etc.) that one finds in a community. These factors increase the legitimacy and success of student projects in the community and make the connection between classroom and community more authentic and enduring.

Exercise 11.1
How Are Your Local Schools Funded?

Find the official webpage of your local school board by going to an Internet search engine (e.g., Google.com, Yahoo.com, etc.) and searching for the name of your town or city and the phrase "school board." Write a one- to two-page paper that answers the following questions:

1. How are your local schools funded?

2. Has funding increased or decreased over the past few years (and why)?

3. How do people become school board members? Are they elected? Are they appointed (and if so, by whom)?

4. Are the school board members nonpartisan, or are they affiliated with a political party? If they are politically affiliated, how many are Democrats? How many are Republicans? Are any independents or are there members with other party affiliations?

5. According to the school board website, what are the key challenges facing your local public schools? How does the board claim to be addressing these challenges?

(Continued)

(Continued)

6. Now, imagine that you are an official member of the school board. What two key challenges do you want to focus on? These may be similar to the ones that the board has identified, or they may be your own, based on the research you have done. Outline these challenges and then describe several ideas you would suggest to address them. For at least one of these suggestions for change, elaborate on your plan and discuss in detail how your skills as a sociologist can help create this change.

Exercise 11.2
Education for All?

Read about the Education for All Global Education Monitoring Report movement (https://en.unesco.org/gem-report/). Then, go to https://www.un.org/sustainabledevelopment/wp-content/uploads/2018/09/Goal-4.pdf/ and read the information about inequality in education across the globe. Then, answer the following questions:

1. Why do you think education was established as a basic right under the 1948 Universal Declaration of Human Rights?

2. Why should you care about whether or not people in other nations have access to education?

3. What do you think that you, working with others, can do to promote the goal of universal education by 2015?

Exercise 11.3
How Does Your Representative Vote?

1. Go to the website for the U.S. House of Representatives (http://www.house.gov).

2. Look up your state representative by entering your zip code under "FIND YOUR REPRESENTATIVE" in the top right corner of the page and clicking on the "LOOK UP" button.

3. When the page opens, click on the representative's name.

4. Find information on the positions your representative is taking on five issues.

5. Answer the following questions:

   a. Do you agree with the votes? (In other words, do they represent you?) Why or why not?

   b. Will you vote in the next (i) local, (ii) state, and (iii) national elections? Why or why not?

   c. Does reading about how your representative has voted recently make you more or less interested in voting for or against your representative in the next election? Why?

   d. If you do not agree with his or her vote on one or more issues, what actions can you (as a citizen) take to express your dissatisfaction?

   e. What actions can you take toward having your representative vote more to your liking (or continue to vote to your liking) on future issues?

Exercise 11.4
Racial Inequality in America's Schools

1. Read the article "US Education: Still Separate and Unequal" (https://www.usnews.com/news/blogs/data-mine/2015/01/28/us-education-still-separate-and-unequal).

2. Go to the "State of Funding Equity Data Tool" at https://edtrust.org/map/ and spend 10 minutes asking the map for different data.

3. Read "Crossing the Gap" (https://www.tolerance.org/magazine/spring-2009/crossing-the-gap).

(Continued)

(Continued)

4. What did you learn? What surprised you about the racial gap in education? Why did it surprise you? How does this mesh with your ideas about the right of every student to have the opportunity for an equal education? With the U.S. ideology that "education is the key to success?" Write a two-page paper addressing these questions (and other questions that you have). Make sure to specifically and clearly utilize all of the above resources.

Exercise 11.5
The Role of Religion in the Lives of Individuals

Survey 10 people about their religion. In the survey, include questions about (a) their religious affiliation (if they have one), (b) how often they go to religious services, and (c) if they think their belief in God or a Higher Power guides their actions during an average day.

Then interview them, asking them to provide examples to illustrate their answers. Compare the results of your surveys and interviews. Were the findings consistent? Why or why not? How might a symbolic interactionist interpret your findings? Why?

Exercise 11.6
Secularism, Religion, and Democracy

Listen to the following stories:

"Hijab Bridges Faith and Fashion for US Muslim Women" at http://fashionfightingfamine.weebly.com/blog/npr-hijab-bridges-faith-and-fashion-for-us-muslim-women

"A Tale of Two U.S. Muslim Women: To Cover or Not?" at http://www.npr.org/templates/story/story.php?storyId=6556263

"Debating the Burqa: Sarkozy Proposes Ban" at http://www.npr.org/templates/story/story.php?storyId=106198806

Then, answer the following questions:

1. What do you believe are the best arguments for allowing women to wear (a) a hijab or (b) a burqa?

2. Which of the arguments you have laid out are the most sociological (taking into account the impact on society as well as on individuals)? Why?

3. Which of the arguments do you agree with most strongly? Why?

DISCUSSION QUESTIONS

1. Before you read this chapter, did you know the names of your U.S. senators and representative? Did you know their party affiliations? Why or why not? Why is it important to know who they are and how they vote?

2. What do you think the purpose of public schools should be? Do you think your education fulfilled that purpose? Why or why not? Did you go to public schools to get your primary and secondary education? Why or why not?

3. Do you think religious symbols should be allowed in public schools in the United States? Why or why not?

4. Do you ever wonder why the word *God* is used so often by public officials in their public statements? Discuss why you think this happens.

5. Do you think students in U.S. public schools should be required to "pledge allegiance to the flag of the United States of America and to the Republic for which it stands, one nation under God, indivisible, with liberty and justice for all"? Why or why not?

6. What role do you think the federal government should play in improving the public education system in the United States? Why? If you think funding for schools should be increased, from where do you think the money should come? Why?

7. Describe the relationship between the economic, family, and educational institutions in your own life. What do you think would happen to you if

you did not have access to a college education? How would that lack of access affect your economic position and your ability to support a family?

SUGGESTIONS FOR SPECIFIC ACTIONS

1. Go to the website http://www.house.gov and find one issue up for debate in the House of Representatives that interests you. Research the issue and write a letter to your representative to encourage her/him/them to vote your way.

2. Attend two different types of religious services, from different religious affiliations, in your local community. Observe and compare what kinds of people are in positions of authority (their gender, race, and age). Write a paper analyzing the dynamics of gender, race, and age in religion.

3. Research an organization that is working on a social justice issue related to inequality in education. Write a two-page essay outlining the scope of the issue, why the organization is addressing it, and how it is carrying out the campaign (i.e., what methods it is using). Include at least one paragraph analyzing the power that groups can have in organizing citizens toward a more just society. If the organization is religious based, include at least one paragraph analyzing the power that religious groups can have in organizing citizens toward a more just society and an additional paragraph that discusses the possible negative consequences of religiously motivated social justice. Finally, determine how you can help with the campaign and follow through with at least one action to do so.

Please go to this book's website at http://study.sagepub.com/white6e to find further civic engagement opportunities, resources, peer-reviewed articles, and updated web links related to this chapter.

NOTES

1. The pledge, originally written in 1892, was amended in 1923, 1924, and 1954. The phrase "under God" was the most recent addition, added during the Cold War in part to distinguish the United States from the USSR, which had outlawed religious practice.
2. See Gary Becker's article "Human Capital and Poverty" at the Acton Institute website (http://www.acton.org/pub/religion-liberty/volume-8-number-1/human-capital-and-poverty).
3. See the federal government's No Child Left Behind Act website (http://www.ed.gov/nclb/landing.jhtml).

4. A U.S. district court ruled in 2005 that the school board's decision to mandate the teaching of intelligent design violated the constitutional separation between church and state (see Associated Press, 2005).
5. The video for Schoolhouse Rock Constitution can be found at https://www.youtube.com/watch?v=Gv3cSmPSmjY
6. The government's failure to fulfill this responsibility during and after Hurricane Katrina in 2005 was (and continues to be) painfully displayed. Hundreds of Americans lost their lives, and thousands more lost their homes, livelihoods, and neighborhoods, because of the breach that occurred in the levees protecting New Orleans.
7. It is important to note that Marx was not exposed to religions that actively promoted social justice, as many have done in the past century. In fact, in recent years, some religious leaders (such as Liberation Theologians) have been accused of being Marxists!
8. The full text can of Marx's "A contribution to the critique of Hegel's Philosophy of Right" can be found at https://www.marxists.org/archive/marx/works/1843/critique-hpr/intro.htm

REFERENCES

Abbasov, S. (2011, January 6). *Azerbaijan: Hijab ban in schools fuels debate in Baku on role of Islam* (EURASIANET.org). Retrieved from http://www.eurasianet.org/node/62670

Associated Press. (2005, December 20). Judge rules against intelligent design. Retrieved from MSNBC website at http://www.msnbc.msn.com/id/10545387

Bakalar, N. (2011, February 7). On evolution, biology teachers stray from lesson plan. *The New York Times*. Retrieved from http://www.nytimes.com/2011/02/08/science/08creationism.html?_r=0

Bottomore, T. (Trans. & Ed.). (1964). *Karl Marx: Early writings*. New York, NY: McGraw-Hill.

Common Core State Standards Initiative. (2018). Standards in your state. Retrieved from http://www.corestandards.org/standards-in-your-state/

Dvorsky, G. (2013, January 15). How 19-year-old activist Zack Kopplin is making life hell for Louisiana's creationists. *Io9*. Retrieved from http://io9.com/5976112/how-19-year-old-activist-zack-kopplin-is-making-life-hell-for-louisianas-creationists

Jansen, J. (2011, December 23). *The civic and community engagement of religiously active Americans* (Pew Internet & American Life Project). Retrieved from http://www.pewinternet.org/Reports/2011/Social-side-of-religious.aspx?src=prc-headline

Jefferson, T. (1802, January 1). *Jefferson's letter to the Danbury Baptists* [Letter]. Retrieved from https://www.loc.gov/loc/lcib/9806/danpre.html

Johnston, J. B. (2013). Religion and volunteering over the adult life course. *Journal for the Scientific Study of Religion, 52*(4), 733–752.

Kingkade, T. (2013, August 1). Ball State University bans teaching intelligent design in science classes. *The Huffington Post*. Retrieved from http://www.huffington post.com/2013/08/01/ball-state-intelligent-design_n_3688857.html

Marx, K. (1844, February). A contribution to the critique of Hegel's Philosophy of Right. Retrieved from https://www.marxists.org/archive/marx/works/1843/critique-hpr/intro.htm

National Assessment of Educational Progress. (2018). The nation's report card. Retrieved from https://www.nationsreportcard.gov/

National Center for Education Statistics. (2018, April). The condition of education. Retrieved from https://nces.ed.gov/programs/coe/indicator_cma.asp

Newport, F. (2017). 2017 update on Americans and religion. Retrieved from Gallup,Inc.websiteathttps://news.gallup.com/poll/224642/2017-update-americans-religion.aspx

Organisation for Economic Cooperation and Development. (2013). *Skilled for life? Key findings from the Survey of Adult Skills*. Retrieved from http://skills.oecd.org/SkillsOutlook_2013_KeyFindings.pdf

Organisation for Economic Co-operation and Development. (2016). *Skills matter: Further results from the Survey of Adult Skills*. Paris, France: OECD Publishing.

Organisation for Economic Co-operation and Development. (2018, June). *OECD economic surveys: United States*. Paris, France: OECD Publishing.

Pew Research Center. (2014). Religious landscape study. Retrieved from Pew Research Center website at http://www.pewforum.org/religious-landscape-study/age-distribution/

Pew Research Center. (2015, November 3). U.S. public becoming less religious. Retrieved from http://www.pewforum.org/2015/11/03/chapter-4-social-and-political-attitudes/

Quenqua, D. (2013, September 2). Young students against bad science. *The New York Times*. Retrieved from http://www.nytimes.com/2013/09/03/science/young-and-against-bad-science.html

Reuters. (2013, October 8). Turkey lifts decades-old ban on Islamic head scarf. *The New York Times*. Retrieved from http://in.reuters.com/article/2013/10/08/turkey-headscarf-ban-idINDEE99706920131008

Talley, J. (2011, April 19). *Banning hijab: Anti-veil laws around the world* (NakedLaw). Retrieved from http://nakedlaw.avvo.com/2011/04/banning-hijab-anti-veil-laws-around-the-world

United States Elections Project. (2018, December 14). 2018 November general election turnout rates. Retrieved from http://www.electproject.org/2018g

U.S. Department of Education. (n.d.). *Elementary and secondary education: ESEA flexibility*. Retrieved from http://www2.ed.gov/policy/elsec/guid/esea-flexibility/index.html

U.S. Department of Education. (2011, September 23). *Obama administration sets high bar for flexibility from No Child Left Behind in order to advance equity and support reform*. Retrieved from http://www.ed.gov/news/press-releases/obama-administration-sets-high-bar-flexibility-no-child-left-behind-order-advance

Wallace, G. (2016, November 30). Voter turnout at 20-year low in 2016. Retrieved from https://www-m.cnn.com/2016/11/11/politics/popular-vote-turnout-2016/index.html

WarnersRetro TV ch2. (2014, March 27). *Schoolhouse Rock the Constitution* [Video file]. Retrieved from https://www.youtube.com/watch?v=Gv3cSmPSmjY

Weber, M. (1970). *From Max Weber* (H. H. Gerth & C. W. Mills, Eds.). Oxford, England: Routledge.

12

The Engaged Sociologist in Action

Congratulations! After reading this book and completing the exercises in the earlier chapters, you have no doubt developed a sociological eye and can make use of your sociological imagination. Now there's no turning back. With a toolbox packed full with *sociological* tools, you cannot help but perceive social patterns that affect all of us but are often unnoticed by those without your sociological training. The best news is that you have also acquired the means to be an effective agent of social change!

Through this sociology class (and perhaps through other means as well), you have developed the skills you need to become an informed and effective citizen who can help shape our society. But as any good superhero knows, with great power comes responsibility. Your sociological training allows you to understand how society works. It is your obligation now to use that knowledge to influence society in ways that will make it better. And just as a carpenter must not only purchase tools but also learn how to use them properly, so must a sociologist in action learn how to use his or her tools. So now it is time for you to practice using your sociological tools to research and address social issues by completing one of the projects we outline below.

The Basic Steps of Social Science Research

No matter what topic you choose to research, you will need to remember and to follow the basic steps of all social science research:

1. Choose a research topic.

2. Find out what other researchers have discovered about that topic.

3. Choose a methodology (how you will collect your data).

4. Collect and analyze your data.

5. Relate your findings to those of other researchers.

6. Do something with your findings!

Select one of the projects below. Then, use the detailed research directions that follow to plan and carry out your chosen project.

Project 12.1 Civic Engagement and Higher Education

In recent years, many educators and leaders in society have talked and written about the obligation of colleges and universities to educate citizens.[1] However, it is still not clear (a) how many *students* think civic engagement should be a part of college education and (b) how many students are actually civically engaged. Your task will be to seek answers to these questions from students on your campus and to devise ways to encourage them to play an active role in shaping society.

Project 12.2 Social Responsibility on Campus

Is your college or university a "socially responsible" institution? In this project, you will (a) use different indicators of social responsibility to measure how socially responsible your school is and (b) use your findings in an effort to make your school more socially responsible.

Some of the questions to consider are the following: Does everyone on campus have equal access to education and other resources? Are campus faculty and staff paid and treated fairly? Do athletes wear or does your campus store sell apparel made in sweatshops? Does your university recycle everything that it can? Has your university invested in any corporations that are cited for human rights violations? Do students of color and faculty of color feel welcome, comfortable, and fairly represented at all levels of the university? Does your school have a center for LGBTQ students, and do LGBTQ students, faculty, and staff feel welcome, comfortable, and fairly represented at all levels of the university? Does your school have an office that works on sustainability and green campus initiatives, and how does your campus compare to others? Are professors of all genders compensated equally? Would students know if your school was not acting in socially responsible ways? If students did know, would they organize to create change?

Project 12.3 Connecting
the Campus to the Community

What are the needs of the community around your college? Are they being met? Does your college or university actively seek input from local leaders on how it might help the community in which it resides? Do student groups and clubs connect the student body with the local community? Is there a chasm (social or status) between the college community and the larger community? Do students have prejudices toward the local community, or does the local community have prejudices toward students? Does your college contribute to the economic well-being of the greater community? This project will examine the social and economic environment of the community of which your school is a part and determine to what extent your school is working to improve the vitality of that community.

Step 1: Research Preparation

A. What You (Think You) Know

Before you start your research, it is important to take stock of what you *think* you know (remember, too, that most of our knowledge is socially constructed) about the topic and *why* you think you know it. This will help focus your thinking and make you aware of some of the potential biases you may have toward the issue. In a one-page paper, (a) describe what you hypothesize (think) will be the answer to your research question and (b) support your hypothesis with examples from your own life, previous research you have done, and your "best guess" sociological analysis (utilizing, of course, your sociological imagination and sociological eye!).

B. Reviewing the Literature

The next step is to examine previous research on your topic. Go to your school library, and locate the databases of academic articles in the social sciences. We recommend Sociological Abstracts, JSTOR, or Academic Search Premier. Conduct several searches for articles using any combination of key words or phrases, such as the following, noting these are just some examples to get you started:

For Project 12.1, "student activism," "student apathy," "college and activism," "civic engagement and college," "educating students," and "student attitudes"

For Project 12.2, "civic engagement and colleges," "student activism," "students and social responsibility," "colleges and social responsibility," and "student movements"

For Project 12.3, "poverty," "hunger," "homelessness," "unemployment," "town and gown," and "campus and community"

If you are having problems with your searches or the search terms are not producing the desired results, please ask a reference librarian at your school for assistance. Reference librarians are amazing resources and can be a huge help to you as you proceed.

Choose four or five recently published articles available in your library that appear relevant to your research question.

Of course, none of the articles will answer the research question completely. However, each may contribute a piece of related data or way of thinking about the question. For each article, note the following:

- Which of your major concepts (e.g., student attitudes, social responsibility, campus and community relations) is discussed?

- What did the researchers find that adds to your knowledge about the question?

Write a two-page summary of the articles you have read. Then, write a one-page paper summarizing what you have learned from the articles that influences your thinking on the question you are addressing. Have your expectations changed at all? Why or why not? If yes, revise your hypothesis.

Step 2: Obtain Your Data

You have developed a research question and conducted some background work. No doubt your thoughts about what you will find from your research have changed since you first thought about the research question. Now, it is time to obtain the data to test whether your present expectations are on target. In each of the three guided exercises below, we present a research design appropriate for each one of the research projects. Be sure to continue with the project you selected when you began Step 1 (i.e., if you chose Project 12.1 in Step 1, you should continue with Project 12.1 throughout the chapter). You can work individually or in small groups, depending on your professor's preference.

Project 12.1 Civic Engagement and Higher Education: Data Collection Exercise

In this exercise, you will survey at least 30 students to determine (a) how interested students at your school are in learning how to become active, effective members of society and (b) the extent to which students on your campus are socially active.

1. Go to this book's website (http://study.sagepub.com/white6e). Select "Research Tools," and download "Survey for Research Project 12.1"; also, download the "Scoring Guide" for Research Project 12.1.

2. Determine where and when to find respondents. Do not just haphazardly hand out forms to people. Rather, select a large general education class that will provide a fairly representative sample of the student body (representing students from different years, majors, races, ethnicities, sexes, etc.). Be sure to secure prior permission from the professor teaching the course before you hand out your survey at the beginning of one of the class meetings.

3. Ask the students to take a 5- to 10-minute survey. Read them the instructions, and then hand out the survey. Instruct the students not to sign their names, and tell them to put the completed surveys face down in a pile in a specified place in the classroom. (That way they know that you will not be able to connect them with their individual responses.)

4. Follow the instructions in the scoring guide to analyze the results. You may want to try grouping your data with those of other students who are also carrying out this project. (This will enable you to have a larger sample on which to base your findings.) Write a three- to four-page paper that presents and analyzes your results and relates them to the previous research you found on the topic.

Project 12.2 Social Responsibility on Campus: Data Collection Exercise

In this exercise, you will interview at least 10 students about how socially responsible they believe your school is. You will ask them questions that address both specific indicators of social responsibility and their overall impression of the social responsibility of the school.

1. Go to this book's website (http://study.sagepub.com/white6e). Select "Research Tools," and download the interview guidelines, the coding guide, and the informed consent statement for Research Project 12.2.

2. Determine where and when to find respondents. Be sure to interview students who are most likely to know (and care) about your topic (e.g., leaders in student government, club officers, members of student-activist groups, student representatives to the board of trustees, etc.).

3. Ask the students you select for permission to interview them. Give them the informed consent sheet to read and sign. Ask permission to tape-record the interview. (If they won't allow you to record, you will need to take detailed notes.)

4. Ask them the questions you devised after reading and following the instructions in the interview guidelines. Start with your own questions, but let the respondent guide the conversation. Do not try to force all of your respondents to answer the same questions in the same way. However, do make sure they each cover all of the questions you have prepared. The point is to get their individual perspectives and then compare them with one another. A good interviewer asks probe questions throughout the interview. These are very short and simple questions such as "Can you expand on that?" or "Interesting. What else might you add?" Probe questions give a cue to the person being interviewed to talk more on the subject while not pushing the interviewee toward your own biases.

5. Follow the instructions in the coding guide to analyze the results. If possible, group your data with those of other students carrying out the same project. (This will enable you to have a larger sample on which to base your findings.)

6. Write a three- to four-page paper that presents and analyzes your results and relates them to the previous research you found on the topic.

Project 12.3 Connecting the Campus to the Community: Data Collection Exercise

Note: For this project, you can collect data through *either* interviews *or* surveys.

Interviews. You will interview leaders in the local community about the relationship between your school and the community.

1. Go to this book's website (http://study.sagepub.com/white6e). Select "Research Tools," and download the interview guidelines, the coding guide, and the informed consent statement for Research Project 12.3.

2. Identify at least three leaders in the local community who would have reason to know the extent and nature of the connection between your school and the community (e.g., the mayor, deputies to the mayor, the city manager, city council members, and directors of local nonprofit agencies). You can find local nonprofit organizations in your area by using World Hunger Year's Grassroots Resources Directory (http://www.whyhunger .org/findfood), which includes organizations that deal with various issues related to poverty, in addition to hunger. Simply fill in the information, and the database will give you a list of organizations in your area. GuideStar also has a database (https:// www.guidestar.org/search). You can type in your state or town and specify the types of organizations you would like to identify. You can also use the database at http://idealist.org to find such organizations.

3. Ask each community leader for permission to do the interview. If he or she agrees, provide a copy of the informed consent sheet to read and sign. Ask permission to tape-record the interview. (If he or she won't allow you to record it, you will need to take detailed notes.)

4. Ask the community leaders the questions you devised after reading and following the instructions in the interview guidelines. Start with your own questions, but let the respondent guide the conversation. Do not try to force all of your respondents to answer the same questions in the same way. However, do make sure they each cover all of the questions you've prepared. The point is to get their individual perspectives and then compare them with one another. A good interviewer asks probe questions throughout the interview. These are very short and simple questions such as "Can you expand on that?" or "Interesting. What else might you add?" Probe questions give a cue to the person being interviewed to talk more on the subject while not pushing the interviewee toward your own biases.

5. Follow the instructions in the coding guide to analyze the results. If possible, group your data with those of other students carrying

out interviews for this project. (This will enable you to have a larger sample on which to base your findings.)

6. Compare your findings with those of your classmates who collected survey data for this project.

7. Write a three- to four-page paper that presents and analyzes your results (and, if possible, the results of your classmates' interviews), relates them to the previous research you found on the topic, and compares them with the findings of your classmates who conducted surveys on this topic.

Surveys. You will survey at least 30 students to measure the (a) attitudes of students toward the local community and (b) degree of interaction between students and the community.

1. Go to this book's website (http://study.sagepub.com/white6e). Select "Research Tools," and download "Survey for Research Project 12.3"; also, download the "Scoring Guide" for Research Project 12.3.

2. Determine where and when to find respondents. Do not just hand out forms to people haphazardly. Select a large general education class that will provide a fairly representative sample of the student body. Be sure to secure prior permission from the professor teaching the course before handing out your survey at the beginning of one of the class meetings.

3. Ask the students to take a 5- to 10-minute survey. Read them the instructions, and then hand out the survey. Instruct the students not to sign their names, and tell them to put the completed surveys face down in a pile in a specified place in the classroom. (That way they know that you won't be able to connect them with their individual responses.)

4. Follow the instructions in the scoring guide to analyze the results. If possible, group your data with those of other students carrying out the survey for this project. (This will enable you to have a larger sample on which to base your findings.)

5. Compare your findings with those of your classmates who conducted interviews for Research Project 12.3.

6. Write a three- to four-page paper that presents and analyzes your results (and, if possible, the results of your classmates' interviews), relates them to the previous research you found on the topic, and compares them with the findings of your classmates who conducted interviews on this topic.

Step 3: Do Something About It

Now that you have personally investigated an important social issue, you are in a strong position to address it. The following exercises will help you use your sociological knowledge to make a difference in the lives of people in your community.

Project 12.1 Civic Engagement and Higher Education: Civic Engagement Exercise

From your research, you have some data that indicate how many students think civic engagement should be a part of college education and how many students are socially active. For this exercise, you will prepare a *talking points* document designed to change the minds of those students who do not think civic engagement should be a part of college education. You will then organize a campus panel with other students at which you will present your talking points.

A talking points document is a brief summary of relevant issues to help you put things into perspective, to establish a context for your concerns, and to focus on what is most important in your argument. It is also a fact sheet that provides evidence to support your position.

1. Drawing on your knowledge of the topic (through your background reading), list the five most important reasons why students should want civic engagement to be a part of their college education.

2. Using information from your survey data, this book, or published research on the topic, identify the specific reasons why students might not want civic engagement to be a part of college education, and then counter their arguments. (For example, if you think students will say that they don't have time to be socially active, mention the fact that many schools have included civic engagement activities within courses. If students don't understand how they can benefit from becoming socially active, describe to them the connection between social activism and social power, etc.)

3. Write a one-page summary of the talking points that is designed to help you communicate your five points to a group of students.

4. Try it out. Bring your talking points to a few friends or relatives, and ask them if they will let you briefly discuss why you think civic engagement should be a part of the college experience.

Think about their responses, criticisms, and suggestions, and then make changes in your document in ways that will strengthen it.

5. Prepare a five-minute presentation based on what you have learned. Consider the reasons why rational and thoughtful people might not want civic engagement to be a part of college education, and think carefully about what you might say that will address and possibly overcome their hesitation. Two websites— http://www.pbs.org/standarddeviantstv/transcript_public.html and http://www.enotes.com/topics/how-write-speech—provide great guides to help you with putting together and delivering your speech.

6. Work with other students who carried out this project to organize and participate in a student panel presentation for the campus community about the connection between civic engagement and higher education.

Project 12.2 Social Responsibility on Campus: Civic Engagement Exercise

Based on your preparation work, you should have a fairly good idea about what social issues students believe your college or university should be addressing on campus (e.g., sweatshops, right to a living wage, health care benefits, sex discrimination, racism).

1. Write a two-page summary of the issue that most concerns you. Include in it a simple description of (a) what the social issue is, (b) who (which group or groups) is suffering as a result of it, (c) who has the power or authority to change it, and (d) what makes you feel that there is a need to address the social issue in the first place.

2. Rewrite your statement in the form of a two-paragraph letter to the editor of your school paper. Begin the letter with the phrase "We the undersigned . . . ." Work and rework this document until you feel confident that you have included your major points and presented them in a clear and powerful manner. At https://ctb.ku.edu/en/table-of-contents/advocacy/direct-action/ letters-to-editor/main and https://www.amnestyusa.org/files/pdfs/ howtowritelettertotheeditor.pdf, you can find great guides to help you with putting together and delivering your speech.

3. Ask people around campus if they would be willing to join you in signing the letter. Provide a signature list form where they can

sign their names, print their names, and (optionally) identify themselves (student, professor, staff, etc.). Also, ask them if they wish to become more active in resolving the injustice. If they do, take down their contact information (including their phone numbers and e-mail addresses).

4. If you find that most people do not agree with your letter, stop and rethink it. Have you represented the issue fairly? What are their objections? Are they right? If so, you might want to consider a different issue or a different approach and start again. If you feel that they are wrong, you should consider what you could do differently in your letter that might convince them to join your cause.

5. After you have collected at least 20 signatures (hopefully more!), send the letter to your school newspaper, to the president of your college, and to any other key people on campus whom you have identified as having the power to affect this issue.

6. Call a meeting of the people who want to act on the issue. Discuss what to do, and do it. Devise an action plan and strategy to take action on your issue. On the following websites, you can find some excellent resources compiled by other student groups that will help you write letters, organize meetings, and gain all of the tools you need to work for your issue:

http://www.campusactivism.org/displayresource-471.htm

http://www3.thestar.com/static/PDF/060217_jrjour_guide.pdf

http://www.campusactivism.org/uploads/FireItUp.pdf

You may also want to consult books such as *Take More Action* by Marc and Craig Kielburger and Deepa Shankaran (2008) or *Welcome to the Revolution* by Charles Derber (2017) for some guidelines, tips, and lessons from activists on how to organize effectively. Michael Gecan's book *Going Public* (2004) is also an excellent overview of how to build power through organizing people. Dave Beckworth and Cristina Lopez's *Community Organizing: People Power From the Grassroots* (n.d.) gives the basics of good organizing.

Project 12.3 Connecting the Campus to the Community: Civic Engagement Exercise

Identify a few nonprofit groups in your area that are working on poverty-related issues. There may be an office on campus that can help you find one, such as the community service or service-learning office.

You can also call your local mayor's office and ask if there are community groups that the city or town works with on poverty-related issues, such as hunger, homelessness, and housing. You should also use tools such as World Hunger Year's Grassroots Resources Directory (http://www.whyhunger.org/connect) and GuideStars search engine (https://www.guidestar.org/search). As we noted above, you can simply fill in the information, and these databases will give you a list of organizations in your area.

You can also find such organizations through the Idealist website at http://idealist.org.

1. Visit the organization you have chosen (or visit several if you are having trouble choosing). Speak with the people there, pick up their literature, and look over their website if they have one. Identify and summarize what their basic work is, how it is organized, and what they want from volunteers or interns. Given your knowledge of the issue, make note of how well the organization's work relates to your area of interest.

2. Get to know the organization better. Attend a meeting, or volunteer for an event. Speak with other volunteers about their experiences.

3. If you are excited by the work of the organization, get a group of students together to volunteer their time and talents to work with the people there. Tell the other potential volunteers what you know about the issue and what this organization is doing. Try to get a meaningful sense of what kind of commitment other students are willing to make so that you don't promise too much, and then offer your group's services to the organization as volunteers.

Two notes of caution should be added before you recruit too many volunteers. First, make sure that you have spoken to the people coordinating the efforts of the organization to find out what they need. They may need help with staffing, clerical work, building repairs, fund-raising, or a variety of other tasks. Or, they may not need help now or at least not the type of help your volunteer group will be able to provide. It is very important that they are part of your process! Second, many organizations prefer to start their volunteers off with clerical tasks, like faxing or filing. The point of this exercise is to get your hands dirty and do something about an ongoing social problem, even if the work you are doing may seem mundane. However, make sure you have a clear sense of what the organization wants from you before you commit to doing it.

Sociologist in Action

You!

(Place a brief biography here and a description of the class project you just completed or any other social change exercise you undertook as part of this course.)

You've now joined the ranks of the other "sociologists in action" highlighted in this book! Like their work, your efforts can inspire others to become knowledgeable, engaged, and effective citizens. Although these sociologists may not look like superheroes, they are, as they have all managed to harness the power of sociology to make society better. Congratulations on becoming one of them! Please e-mail us your Sociologist in Action piece (jmwhitebentley@gmail.com), and let us know if we can use it on our website or in future editions of *The Engaged Sociologist: Connecting the Classroom to the Community* or our other books, *Sociologists in Action: Sociology, Social Change, and Social Justice* and *Sociologists in Action on Inequalities: Race, Class, Gender, and Sexuality*. Professors, we'd love to hear your Sociologist in Action stories too!

Conclusion

You now have sociological tools, skills, and knowledge. You do not have to be a professional sociologist to use them. Just keep your eyes open to how society is working (the sociological eye), make connections between personal troubles and public issues (the sociological imagination), and always look for multiple perspectives. Ask questions, and find your own answers. Question your answers, and always dig deeper. Change is always happening all around us. Whether or not we choose to help direct it is up to us. We don't know all of the social problems or issues you will confront in your life, but we do know that you can make a difference. The anthropologist Margaret Mead once said, "Never doubt that a small group of thoughtful, committed citizens can change the world. Indeed, it's the only thing that ever has." Using your sociological tools, *you* can help change the world!

NOTE

1. The Campus Compact and American Democracy Project efforts are a direct result of this growing realization among academic and government leaders. You can find many sources and links to articles about this topic on the websites for these organizations (http://www.compact.org and http://www.aascu.org/programs/adp, respectively).

REFERENCES

Beckworth, D., & Lopez, C. (n.d.). *Community organizing: People power from the grassroots*. Retrieved from http://comm-org.wisc.edu/papers97/beckwith.htm

Derber, C. (2017). *Welcome to the revolution: Universalizing resistance for social justice and democracy in perilous times*. Oxfordshire, England: Taylor & Francis.

Gecan, M. (2004). *Going public*. New York, NY: Random House.

Kielburger, M., Kielburger, C., & Shankaran, D. (2008). *Take more action*. Toronto, Ontario, Canada: Thomson/Nelson.

Index

Gender socialization
   cultural differences, 56
   impact of, 177, 179, 183–184,
      185–186, 200 (n5), 208
   intersectionality and, 178
Generalized other, 76–77
Genocide
   in Darfur, 62
   Holocaust, 20, 161
   in Rwanda, 187
Gentlemen's Agreement of 1907, 152
Germany, Nazi, 20, 161
Ginsburg, Ruth Bader, 2
Glass-Steagall Act, 127
Global warming. See Climate change
Glogowski, Brianne, 82–84
Gora, Jo Ann, 228–229
Gore, Al, 105
Government institutions
   constitutional framework, 229–230
   education funding and policies,
      227–228
   functions, 230
   local, 228, 230
   need for, 205
   religion and, 226, 232–233
   taxes, 41–42
Great Recession (2008 to 2009),
      127–128, 153

Habitat for Humanity, 213
Harvard Business School, 183
Harvard University, 24–25
Hate crimes, 38–39, 59, 181
Hate groups, 59
Health, toxic chemicals and, 64–65
Health insurance, 210
Higher education. See Education;
   Student activism
Hijabs, 232–233
Hispanics
   economic inequality, 156–157
   gender gaps, 179
   immigrants, 153
   organizations, 135
   population, 154
   poverty rate, 133

racial categories, 153–154
racial profiling of, 100
wages, 208
See also Ethnicity
Hitler, Adolf, 20
Hochschild, Arlie R., 183
Holland, Curtis, 7–9
Homeless individuals, 134,
   136–137, 234
Homosexuals. See LGTBQ individuals
Huerta, Maria, 83
Hull House, 22, 23–24, 25
Hussein, Saddam, 27

IAF. See Industrial Areas Foundation
Id, 77
Immigrants
   Asian, 151, 152, 153
   dehumanizing language, 80
   Mexican, 80, 152, 153–154
Immigration
   cultural influences, 58
   laws and policies, 151–153, 173 (n2)
   push and pull factors, 152–153
   race/ethnicity and, 151–153
   in Western Europe, 161
Immigration Act of 1965, 152
Incomes. See Economic
   inequality; Wages
Industrial Areas Foundation (IAF), 136
Inequality
   class conflict and, 17
   Durkheim on, 21
   economic, 128–129, 129 (figure),
      131, 132, 156–157, 184–187,
      210–212
   educational, 227
   external, 21
   gender-based, 182–187
   global, 210–211
   internal, 21
   racial, 154–158
   Weber on, 19
   See also Social stratification
Information, critical consumption of, 6
Institutional review boards, 53 (n6)
Institutions. See Social institutions

on social media, 62–63
social stratification and, 133–135
See also Civic engagement; Elections
Pollution. *See* Environmental issues
Popular culture, 180
See also Television
Poverty, 131, 133, 185, 211, 212
See also Economic inequality
Power
corporate, 98, 130, 131
gender and, 182–187
Weber on, 19
Power elite theories, 130–131
Preventing War, Promoting Peace, 131
Prejudice, 151, 155
Princeton Review, 61–62, 108–109
Prisons, 233–234
Proletariat, 18, 130

Questions
interview, 42
survey, 41–42, 44
See also Research questions

Race
biracial individuals, 154
color-blind ideology, 158–160
critical race theory, 25, 156
defined, 154
Du Bois on, 24–25
groups in United States,
154, 155 (table)
hierarchies, 154–156
immigration and, 151–153
interracial marriages, 101, 154
intersectionality, 156, 178
media depictions, 157–158
segregation, 25, 101–102, 157
social class and, 133
social construction of, 153–158
Race-centered theorists, 155–156
Racial discrimination, 25, 151, 156,
157–158, 159–160
Racial minorities, 154
See also African Americans; Asian
Americans; Native Americans
Racial prejudice, 151, 155

Racial profiling, 100
Racism
defined, 151
environmental, 39, 160
as global issue, 160–161
immigration policies and, 151–152
internalized, 77
jokes and, 39–40
persistence, 155, 158–160
Reagan, Ronald, 2
Reed, Brian J., 27
Reiman, Jeffrey, 98
Reliability, 41, 44
Religion
conflict perspective, 231–232
evolution/intelligent design debates,
228–229
extremist groups, 59, 72 (n3)
freedom of, 232–233
functionalist view, 231
government and, 226, 232–233
organized, 206, 231
socialization role, 82
symbolic interactionist view, 26, 232
See also Christians; Islam
Republicans, 63–64
Research
literature reviews, 246–247
methods, 6, 27, 41–44
objectivity, 44–45
observational, 42–43, 53 (n6)
projects, 245–246
reliability, 41, 44
steps, 37, 40, 246–254
validity, 41, 44
Research questions
how do you know, 40–44
now what, 37–39
so what, 39–40
Rudolph, Eric Robert, 72 (n3)
Rwanda, 187–188

St. Patrick's parish, Brockton,
Massachusetts, 97
Same sex marriage, 2–3, 179–181, 209
Saudi Arabia, religious dress, 233
SBC. *See* South Bronx Churches

Transgender individuals, 178, 181, 200
  (nn2–3)
Trump, Donald J.
  anti-immigrant language, 80
  election (2016), 61, 134, 135,
    158, 179
Trump administration
  climate change policies, 103, 107
  environmental policies, 103, 160
  hate crime increase during, 59
  immigration policies, 153
Turkey, laws on religious dress, 233
Twenty-Sixth Amendment, 205

Unemployment, 133, 152–153, 157
UnidosUS/National Council of
  la Raza, 135
Unions, 134, 212
United Students Against Sweatshops
  (USAS), 6
Universities. *See* Education; Student
  activism
University of Chicago, 22
University of Wisconsin Oshkosh, 61
USA Hockey, 178
USAS. *See* United Students Against
  Sweatshops

Validity, 41, 44
Values, 55–56, 57, 58, 60
Van Hook, Michele, 82–84
Voting rights, of women, 184
  See also Elections

Wages, 134, 136, 147 (n4), 208, 212
Warren, Joshua, 7–9
Wealth. *See* Economic inequality;
  Social class
Weber, Max, 17, 19–20, 130, 210, 226
WE Charity, 214
Western Europe. *See* Europe
White, Jonathan, 131
White-collar workers, 133, 147 (nn3–4)
White privilege, 157–158, 159
Whites
  color-blind ideology and, 159
  economic inequality, 156–157

population, 154
  See also Race
Wilbur, Bria, 7–9
William Paterson University,
  82–84
Women
  athletes, 178
  education, 59, 185
  employment, 147 (n3), 181–182,
    183, 185–186, 187–188, 209
  Supreme Court justices, 2
  voting rights, 184
  See also Gender
Woods, Tim, 233–235
Worker Rights Consortium (WRC), 6,
  8, 61, 212
Workers
  blue-collar, 133, 147 (nn3–4)
  education of, 226, 227
  false consciousness, 18
  food insecurity, 212
  Marx on, 17–18
  proletariat, 18, 130
  union membership, 134, 212
  wages, 134, 136, 147 (n4),
    208, 212
  white-collar, 133, 147 (nn3–4)
  See also Employment;
    Unemployment
Working class, 18, 23, 130, 133,
  134, 135
World Bank, 107, 211
World Trade Organization (WTO), 211
WRC. *See* Worker Rights Consortium
WTO. *See* World Trade Organization

Yancey, George, 154
Yang, Donna, 82–84
Yousafzai, Malala, 59
Youth
  mentoring programs, 99
  voter turnout, 61
  voting age, 205
  See also Student activism
Yum Brands, 61

Zimmerman, George, 158–159

About the Authors

Jonathan M. White, PhD, is director of the Bentley Service-Learning and Civic Engagement Center and associate professor of sociology at Bentley University in Massachusetts. His primary areas of specialization are inequality, poverty, globalization, human rights, and public sociology. Dr. White has received numerous teaching and humanitarian awards. He was founding director of Sports for Hunger and the Hunger Resource Center and created the Halloween for Hunger (now WE Scare Hunger), WE are Silent, and Pass-the-Fast campaigns. He serves as chair of the U.S. board of directors for the WE movement and on the board of directors for Peace Through Youth, the Graduation Pledge Alliance, Waltham Family Schools, SEED, and on the Global Education Council for the Millennium Campus Network. Additionally, Dr. White is an organizational consultant to a variety of small and large, local, national, and international nonprofits on strategic planning and with universities on building service-learning and civic engagement programs. Dr. White has authored articles and book chapters in the fields of inequality and globalization. His work on public sociology includes coediting (with Shelley White and Kathleen Korgen) *Sociologists in Action: Sociology, Social Change, and Social Justice* (2014) and *Sociologists in Action on Inequalities: Race, Class, Gender, and Sexuality* (2015). He is also currently writing a book titled *Hungry to Be Heard: Voices From a Malnourished America*.

Shelley K. White, PhD, MPH, is director of the Master of Public Health Program in Health Equity and associate professor of sociology and public health at Simmons University. Her teaching and research focus on global and domestic health inequities, health policy, HIV/AIDS, globalization and trade, and human rights and social justice. Her recent publications appear in the *Journal of Human Rights Practice*; *Health, Risk and Society*; *Medicine, Conflict and Survival*; and the *American Journal of Public Health*, and she proudly co-edited the *Sociologists in Action* series with Jonathan White and Kathleen Korgen and *Preventing War and Promoting Peace: A Guide for Health Professionals* with William Wiist. As a scholar-activist, she is committed to engaging with social justice movements and organizations, chairing and serving on the boards of directors for some, including Human Rights Cities: Boston and Beyond, the People's Health Movement, the WE Movement, single payer movements, local anti-gentrification movements, the Public Health Working Group on Primary Prevention of War, the Campaign Against Racism through SocMed, and the Social Medicine Consortium.

Jonathan and Shelley live in Waltham, Massachusetts, and are the very proud uncle and aunt of their 13 nieces and nephews, Jarred, Kyle, Tyler, Arielle, Cameron, Brianna, Mikayla, Joshua, Jack, Logan, Tyler, Joey, and Brookelyn.